WILDLIFE PROTECTION, DESTRUCTION AND EXTINCTION

IMPACTS OF THE GULF OIL SPILL ON FISHING AND WILDLIFE

Wildlife Protection, Destruction and Extinction

Additional books in this series can be found on Nova's website under the Series tab.

Additional E-books in this series can be found on Nova's website under the E-books tab.

Fish, Fishing and Fisheries

Additional books in this series can be found on Nova's website under the Series tab.

Additional E-books in this series can be found on Nova's website under the E-books tab.

WILDLIFE PROTECTION, DESTRUCTION AND EXTINCTION

IMPACTS OF THE GULF OIL SPILL ON FISHING AND WILDLIFE

CARLOS SCARNATI
AND
EDUARDO POPEO
EDITORS

Nova Science Publishers, Inc.
New York

For permission to use material from this book please contact us:
Telephone 631-231-7269; Fax 631-231-8175
Web Site: http://www.novapublishers.com

Library of Congress Cataloging-in-Publication Data

Impacts of the Gulf oil spill on fishing and wildlife / editors: Carlos Scarnati and Eduardo Popeo.
p. cm.
Includes index.
ISBN 978-1-61470-971-8 (hardcover)
1. BP Deepwater Horizon Explosion and Oil Spill, 2010. 2. Oil spills--Environmental aspects--Mexico, Gulf of. 3. Fisheries--Mexico, Gulf of. 4. Marine mammals--Effect of oil spills on--Mexico, Gulf of. 5. Sea turtles--Effect of oil spills on--Mexico, Gulf of. I. Scarnati, Carlos. II. Popeo, Eduardo.
GC1221.I45 2011
338.3'7270916364--dc23

2011028

Published by Nova Science Publishers, Inc. ✦ New York

CONTENTS

PREFACE

On April 20th, 2010 the Deepwater Horizon oil drilling rig was destroyed by an explosion and fire, and the oil well began releasing oil into the Gulf of Mexico. The oil spill caused significant economic harm to the Gulf fishing industry because of fishery closures and consumer concerns related to the safety of Gulf seafood. Intermediate and long-term concerns are related to impacts on marine populations and degradation of fisheries habitat necessary for spawning, development of early life stages and growth. This book presents information related to damages caused by the oil spill to Gulf fisheries and wildlife and efforts to mitigate these damages. Many uncertainties exist because of the complexity and scale of Gulf fisheries and ecosystems that have been affected by the oil spill. Direct and indirect damages to fisheries and the Gulf environment are still being assessed and these efforts are likely to continue for years to come.

Chapter 1- On April 20, 2010, the Deepwater Horizon oil drilling rig was destroyed by an explosion and fire, and the oil well began releasing oil into the Gulf of Mexico. The oil spill caused significant economic harm to the Gulf fishing industry because of fishery closures and consumer concerns related to the safety of Gulf seafood. Intermediate and long-term concerns are related to impacts on marine populations and degradation of fisheries habitat necessary for spawning, development of early life stages, and growth.

Chapter 2- The explosion of the Deepwater Horizon drilling rig in the Gulf of Mexico on April 20, 2010, and the resulting oil spill began a cascade of effects on the coastal areas of the Gulf and on the wealth of species that inhabit those areas, effects that continue even though oil stopped flowing in July and the well was sealed in September. Gulf coast wetlands, like those elsewhere, have value for water quality, flood control, shoreline protection, and recreation. They serve as nurseries for many species, including fish and shellfish of commercial significance, waterfowl, and a host of resident and migratory species. They also have cultural importance to the people of the Gulf. The effects of the spill come on top of historic wetland losses due to subsidence, drainage, and saltwater intrusion, along with rising sea levels, coastal erosion, and global climate change.

Chapter 3- “The oil didn’t do us any good but it sure didn’t hurt Florida fisheries,” says Lee Schlesinger, spokesman for the Florida Fish and Wildlife Conservation Commission. “Still, the idea that our seafood may not be safe is stuck in people’s minds. People saw the oil gushing every night on CBS and they saw that Florida was on the same map. You can’t truth your way out of it.”[1] Florida temporarily closed a small portion of state waters to fishing last summer due to the BP oil spill but Schlesinger says the closure had little *direct* effect on the Florida fishing industry: The closure was brief and did not occur in an area vital to commercial fisheries. Gulf seafood has since been subjected to months of testing pursuant to a seafood safety protocol established by state and federal agencies in response to the spill. So far, no sample has tested positive for contamination near the thresholds considered harmful to human health.

Chapter 4- Disaster relief may be provided by the federal government to assist the fishing industry when it is affected by a commercial fishery failure. A commercial fishery failure occurs when fishermen endure economic hardships resulting from fish population declines or other disruptions to the fishery. The Department of Commerce can provide disaster assistance under Sections 308(b) and 308(d) of the Interjurisdictional Fisheries Act (16 U.S.C. § 4107), as amended, and Sections 312(a) and 315 of the Magnuson-Stevens Fishery Conservation and Management Act (16 U.S.C § 1861). The National Marine Fisheries Service plays a central role in determining whether a commercial fishery failure has occurred and in allocating federal funding to states and affected fishing communities. Congress plays a pivotal role by appropriating funds and providing oversight of the process.

Chapter 5- Direct contact with of petroleum compounds or dispersants with skin may cause skin irritation, chemical burns, and infections. Inhalation of volatile petroleum compounds or dispersants may irritate or injure the respiratory tract, which may lead to inflammation or pneumonia. Ingestion of petroleum compounds may cause injury to the gastrointestinal tract, which may affect the animals’ ability to absorb or digest foods. Absorption of petroleum compounds or dispersants may damage liver, kidney, and brain function as well as causing anemia and immune suppression. Long term chronic effects such as decreased survival and lowered reproductive success may occur. The Marine Mammal and Sea Turtle Stranding Networks have protocols and procedures in place for responding to live animals that are exposed to oil spills, and animals brought into rehabilitation facilities are provided veterinary care to remove oil and treat any related health effects.

Chapter 6- Sea turtle stranding responders, working under the guidance of the Wildlife Branch Unit of the Unified Command, which includes several NOAA experts, are responding to dead and live sea turtle strandings and offshore teams are searching for and rescuing oiled sea turtles from the spill area.

Chapter 7- Healthy coral reefs are among the most biologically diverse and economically valuable ecosystems on earth, providing valuable and vital ecosystem services.

Coral ecosystems are a source of food for millions; protect coastlines from storms and erosion; provide habitat, spawning and nursery grounds for economically important fish species; provide jobs and income to local economies from fishing, recreation, and tourism; are a source of new medicines, and are hot spots of marine biodiversity. They also are of great cultural importance in many regions around the world, particularly Polynesia.

In: Impacts of the Gulf Oil Spill on Fishing … ISBN: 978-1-61470-971-8
Editors: C. Scarnati and E. Popeo

Chapter 1

THE DEEPWATER HORIZON OIL SPILL AND THE GULF OF MEXICO FISHING INDUSTRY*

Harold F. Upton

SUMMARY

On April 20, 2010, the Deepwater Horizon oil drilling rig was destroyed by an explosion and fire, and the oil well began releasing oil into the Gulf of Mexico. The oil spill caused significant economic harm to the Gulf fishing industry because of fishery closures and consumer concerns related to the safety of Gulf seafood. Intermediate and long-term concerns are related to impacts on marine populations and degradation of fisheries habitat necessary for spawning, development of early life stages, and growth.

The closing and opening of fishing grounds has involved a tradeoff between ensuring public safety and providing fishing opportunities to recreational and commercial fishermen. In addition to public health concerns, uncertainties related to Gulf seafood safety could further compromise the reputation of Gulf seafood. Most areas have been reopened and landings of commercial and recreational species are recovering. For Gulf waters re-opened through November 15, 2010, sensory analyses of seafood samples have found no detectable oil or dispersant odors or flavors, and results of chemical

* This is an edited, reformatted and augmented version of Congressional Research Service Report R41640, dated February 17, 2011.

analyses have been well below levels of concern. However, some scientists and the public remain skeptical of claims that Gulf seafood is safe. This may inhibit the recovery of Gulf recreational and commercial fisheries.

Under the Oil Pollution Act (OPA), harmed individuals and businesses may make claims for economic injuries to the responsible party, in this case BP. Although many in the fishing industry have benefited from their damage claims and associated payments, ongoing issues include the legitimacy of some claims, lack of transparency in the claims review process, eligibility to make a claim, and level of payments. Other assistance to the fishing industry includes BP grants to states, National Oceanic and Atmospheric Administration (NOAA) fishery disaster assistance, the BP Vessels of Opportunity Program, and Small Business Administration efforts.

Environmental restoration of fisheries habitat and Gulf ecosystems would support the long-term recovery and productivity of Gulf fisheries. The federal government's role in restoration is defined in statute by OPA and in NOAA regulations, which require development of a Natural Resource Damage Assessment (NRDA). NRDA restoration plans are currently being developed by state and federal trustees. The Obama Administration also has committed to developing a separate long-term Gulf of Mexico plan to restore the environment, economy, and public health of residents. Implementation of the plan will require sustained funding and a governance structure to oversee and coordinate restoration efforts.

The 112th Congress may continue to conduct oversight of efforts to promote fishing industry recovery, adequate compensation to fishermen and businesses, and Gulf restoration. Ongoing efforts by federal agencies and states to ensure seafood safety and to regain and maintain the reputation of Gulf seafood are the most immediate challenges currently faced by the fishing industry. As the NRDA process moves from the planning to restoration phase, questions may arise regarding the level of the potential settlement and the types of restoration activities identified by the trustees. In contrast to NRDA, three bills have been introduced in the 112th Congress to address elements of the Administration's restoration plan. All three bills would establish a Gulf Coast Ecosystem Restoration Fund and require 80% of any amounts collected by the United States as penalties, settlements, or fines under the Federal Water Pollution Control Act to be deposited into the fund. They would also establish a governing body to distribute funding and coordinate restoration efforts. Potential issues involve the allocation of funds, focus of restoration projects, and coordination with other restoration efforts.

INTRODUCTION

The Deepwater Horizon oil spill has caused significant socioeconomic injuries to the Gulf of Mexico fishing industry. Immediate economic injuries occurred because large areas of federal and state waters in or adjacent to the spill area were closed to fishing as a precautionary measure to ensure the safety of seafood. Perhaps of greater concern are intermediate and long-term harm to Gulf of Mexico ecosystems. Seafood production is dependent on Gulf ecosystems for spawning areas, nurseries, and growth.

Over 84 days beginning on April 20, 2010, the Deepwater Horizon oil well released over 200 million gallons (4.9 million barrels).[1] During the spill, oil was dispersed through the application of 1.8 million gallons of surface and subsurface chemical dispersants. By June 2, 2010, the area of federal waters closed to fishing had grown to its maximum of 88,522 square miles or nearly 37% of federal waters in the Gulf of Mexico.[2] The maximum portions of state waters closed to fishing during the spill were Alabama (40%), Florida (2%), Louisiana (55%), and Mississippi (95%).[3] Survey teams also have documented 1,053 total linear miles of oiled shoreline.[4]

At issue for Congress is whether current efforts will be sufficient to restore the fishery to pre-spill conditions. Congress may continue to conduct oversight of efforts to promote fishing industry recovery and Gulf restoration. Immediate issues are likely to focus on providing financial compensation to fishermen, ensuring seafood safety, and maintaining Gulf seafood markets.

Congress also may consider long-term efforts to restore the Gulf environment and related fishery productivity. In addition to restoration requirements defined in statute by the Oil Pollution Act (OPA, P.L. 101-380) and National Oceanic and Atmospheric Administration (NOAA) Natural Resource Damage Assessment (NRDA) regulations under OPA, the Administration has proposed a long-term plan to restore the Gulf region. The plan would redirect Clean Water Act penalty revenues for this purpose[5] and establish a Gulf Restoration Council to coordinate restoration efforts. Congress may consider the Administration's proposal and related issues such as the allocation and use of restoration funds.

The primary objective of this report is to summarize information related to damages caused by the oil spill to Gulf fisheries and efforts to mitigate these damages. Many uncertainties exist because of the complexity and scale of Gulf fisheries and ecosystems that have been affected by the oil spill. Direct and indirect damages to fisheries and the Gulf environment are still being assessed and these efforts are likely to continue for years to come.

Table 1. Commercial Fisheries Statistics by State in 2008 Landings and revenue in thousands.

State	Total Landings and Revenue		Revenue for Major Commercial Species			
	Landings	Revenue	Shrimp	Menhaden	Oysters	Blue Crab
West Florida	58,643	$162,182	$ 23,265	$ 15	$ 5,473	$ 3,300
Alabama	24,534	$ 44,234	$ 38,355	$ 59	$ 243	$ 1,533
Mississippi	201,822	$ 43,697	$ 17,146	$18,534	$ 6,869	$ 447
Louisiana	915,956	$272,857	$130,623	$45,768	$38,852	$32,185
Texas	72,469	$174,621	$157,182	0	$ 8,835	$ 2,341
Totals	1,273,424	$697,591	$366,571	$64,376	$60,272	$39,806

Sources: National Marine Fisheries Service, *Fisheries of the United States 2009*, Silver Spring, MD, July 2010, p.6. http://www.st.nmfs.noaa.gov/st1/fus/fus09/index.html, http://www.st.nmfs.noaa.gov/st1//fus/fus09/ 03_recreational2009.pdf. NOAA Fisheries Office of Science & Technology, Fisheries Statistics, Queried commercial landings and revenue February 10, 2010, http://www.st.nmfs.noaa.gov/st1/commercial/landings/annual_landings.html.

Commercial and Recreational Fisheries

Other than Alaska, historically, the Gulf region has produced the greatest amount of seafood by volume and value in the United States. In 2008, Gulf commercial fishery[6] landings totaled 1,273 million pounds with a dock-side value of $697 million.[7] When related processor, wholesale, and retail businesses are included, the seafood industry of the Gulf states supported over 213,000 full- and part-time jobs with related income impacts of $5.5 billion.[8] The top commercial species by value were shrimp ($366 million), menhaden ($64 million), oysters ($60 million), and blue crab ($39 million).[9] **Table 1** provides commercial landings and revenue for major species by state in the Gulf of Mexico region.

Recreational fisheries also make significant contributions to the region's economy by supporting businesses such as charters, bait and tackle shops, restaurants, and hotels. In 2008, 5.7 million Gulf recreational fishermen, both visitors and residents, took 24 million fishing trips.[10] In 2008, recreational fishermen spent over $12.5 billion on durable equipment[11] and trips in the Gulf region.[12] Some of the main species targeted by recreational fishermen include snappers, several types of drum, sheepshead, and Spanish mackerel. Although recreational expenditures cannot be directly compared to commercial revenues, these figures provide an indication of the magnitude of economic activity related to these fisheries. Commercial and recreational fisheries are among the main activities supporting the economy and social well-being of many Gulf coastal communities.

IMPACTS OF THE OIL SPILL ON GULF FISHERIES

The Deepwater Horizon oil spill has directly harmed the fishing industry through closures and changes in seafood demand. Large areas of federal and state waters were closed to fishing as a precautionary measure to ensure the safety of seafood. During the closures fisheries landings and associated revenues decreased significantly in central Gulf fishing ports. It is likely that demand for Gulf seafood has decreased because of changes in consumer perceptions related to the spill. The spill also has harmed the Gulf environment, resulting in mortality of organisms, eggs, and early life stages and harm to habitat and other elements of the Gulf ecosystem. The scale and nature of the spill make it difficult for the public or government to quantify these effects.

Impacts on Landings and Markets

Closing and Reopening Areas

The closing and opening of areas has involved a tradeoff between ensuring public safety and providing fishing opportunities to recreational and commercial fishermen. In addition to public health concerns, marketing of oil-tainted products would further compromise the reputation of Gulf seafood. On the other hand, closures directly constrain recreational and commercial fishermen and delays in reopening areas are costly to the fishing industry.

On May 2, 2010, 12 days following the explosion and fire of the Deepwater Horizon, NOAA closed 6,817 square miles of the Gulf of Mexico to commercial and recreational fishing. The closure was implemented to ensure potentially contaminated seafood would not enter markets and pose a risk to human health. The closure grew to include portions of Louisiana, Mississippi, Alabama, and Florida state waters. At the peak of the closure, 88,522 square miles, or nearly 37%, of all federal waters in the Gulf of Mexico were off-limits to fishing.[13] The maximum proportions of state waters closed to fishing during the spill were Alabama (40%), Florida (2%), Louisiana

(55%), and Mississippi (95%).[14] Since the flow of oil from the well-head was stopped in July, most of Louisiana state waters and all of Mississippi, Alabama, and Florida state waters have been re-opened to fishing. As of January 24, 2011, only 1,041 square miles of federal waters immediately surrounding the well-head remain closed to commercial and recreational fishing.[15]

FDA, NOAA, and coastal states established a protocol to determine when areas may be re-opened to fishing.[16] Once areas have been determined to be free of oil from the spill, re-opening has been considered on a species-by-species basis. Seafood samples of the species in question must pass both sensory and chemical analyses to ensure there are no harmful oil residues.[17] For sensory testing, edible portions of the species are tested by a panel of experts who check samples for oil and dispersant odor and taste. If all tested samples for a given site pass the sensory test, additional samples undergo chemical analysis to test for polycyclic aromatic hydrocarbons (PAHs) and dispersants. All seafood samples from an area must pass both tests for the area to be reopened to fishing.

Some have criticized the testing protocols because they believe seafood sampling coverage has been insufficient, the setting of PAH levels of concern should have incorporated additional factors, and the list of toxic substances being tested is too narrow.[18]

NOAA has collected samples from federal waters while state personnel have collected samples from state waters. On August 19, 2010, FDA officials stated the following in congressional testimony:

> To date all samples have passed sensory testing for oil or dispersants and, as with the surveillance sampling, the results of all chemical analyses have shown PAH levels well below the levels of concern, usually by a factor of 100 to 1,000 below those levels, essentially at the same level as were seen before the spill.[19]

For federal waters reopened through November 15, 2010, sensory analyses have found no detectable oil or dispersant odors or flavors, and results of chemical analyses have been well below levels of concern.[20] Further, NOAA and FDA sampling from commercial landing sites and markets have not found seafood contaminated by oil or dispersant. According to the FDA, "fish and shellfish harvested from areas re-opened or unaffected by the oil spill are considered to be safe to eat."[21]

Impacts on Fisheries

The areas affected by the closures are some of the richest fishing grounds in the Gulf for commercial species such as shrimp, menhaden, and oysters. Although many factors influence commercial landings, when compared to the same period in 2009 (January through December), total Gulf landings for all shrimp species in 2010 decreased by 35.6 million pounds (27%).[22]

SEAFOOD SAFETY CONCERNS RELATED TO OIL AND DISPERSANTS

Crude oil contains a mixture of chemicals including polycyclic aromatic hydrocarbons (PAHs) that may accumulate in the tissues of marine organisms. The presence of PAHs can make seafood unfit for human consumption because some PAHs have been shown to be carcinogenic. Petroleum products may also taint seafood with an oily smell and taste. The Food and Drug Administration (FDA) considers tainted fish to be adulterated and does not permit the sale of adulterated foods.[24] Unlike oil, dispersants have a low potential to bio-accumulate in fish because they are quickly diluted in water. Dispersants biodegrade in weeks to months, further decreasing the likelihood that they would remain in ecosystems as some components of crude oil might.[25] With time and distance from the point of application, concentrations decrease rapidly.

The primary concern with dispersants is not necessarily related to their toxicity, but how dispersants may alter the fate of oil in the environment. According to a consensus statement developed by a diverse group of scientists, the use of Corexit dispersants (the type of dispersant used in the Gulf of Mexico) increases the concentration of hydrocarbons in the water column and can increase the toxicity of oil because dispersants can work as a delivery system, allowing dispersed oil to enter organisms more readily.[26] According to a 2005 National Research Council Report,

In many instances where a dispersed plume may come into contact with sensitive water column or benthic organisms or populations, the current understanding of key processes and mechanisms is inadequate to confidently support a decision to apply dispersants.[27]

On the other hand, a May 2010 FDA memorandum concluded that "the available information indicates that dispersants have little or no effect on the accumulation potential of oil contaminants, nor do they themselves accumulate in seafood."[28] There remains considerable uncertainty related to the effects of dispersed oil on marine organisms. The NRC report's recommendations included the setting of priorities for dispersant research, including the need to fund a series of toxicity studies to determine the mechanisms of both acute and sublethal toxicity to organisms from exposure to dispersed oil.

On the state level, shrimp landings decreased by 32% in Louisiana, 60% in Mississippi, 56% in Alabama and nearly 15% in Texas, while increasing by nearly 15% for the Florida west coast. Menhaden landings in Louisiana also decreased by 171 million pounds (17%). The decreases in landings resulted in lower revenues in the harvesting sector. Immediate losses were dependent on target species, location, and alternative fishing opportunities. Landings information for other species and associated revenue information are not available at this time.

In addition to impacts resulting directly from oil, the oyster industry has been harmed by efforts to protect Louisiana estuaries from oil intrusion. Following the oil spill, the state of Louisiana released freshwater from the Mississippi River into estuaries in greater amounts than usual in an attempt to keep oil from reaching Louisiana's estuaries. The strategy may have had limited success in keeping oil offshore, but with unintended consequences. The freshwater releases decreased salinity on the oyster grounds below the level that oysters can tolerate and resulted in significant mortality of oysters, by some estimates 50% of Louisiana's annual oyster crop. Some of the oyster industry's immediate concerns are related to documenting damage caused by the spill and freshwater diversion and effects of the spill on consumer perceptions and oyster markets. Prior to the oil spill, long-term restoration proposals to divert the Mississippi River for wetlands restoration raised concerns because of likely impacts on the productivity of oyster grounds. Some in the oyster industry have questioned whether current restoration efforts may need to consider retiring oyster grounds and developing new areas that are less likely to be affected by river diversions associated with restoration activities.[23]

Impacts on Markets

The Gulf oil spill has affected both the supply of Gulf seafood and the demand for Gulf seafood. Fishery closures constrained harvesters and disrupted seafood supplies for the region's processors, distributors, and buyers. The disruption of seafood supplies resulted in the loss of some of the region's seafood markets. Impacts on specific markets vary depending on the magnitude of changes in supply from the harvest sector, whether the market is local, regional, or national, and the availability of alternative supplies. The disruption in Gulf supplies is likely to have induced buyers to use substitutes such as products from other regions or imports. Many in the Gulf seafood industry are concerned that once seafood buyers switch seafood suppliers, it may be difficult to regain markets.

Many in the Gulf seafood industry fear it also will be difficult to regain consumer trust in their products. A study conducted by MRops, a marketing research company commissioned by the Louisiana Seafood Promotion Board, reported that 70% of consumers polled expressed some level of concern about seafood safety following the Gulf oil spill and 23% have reduced their consumption of seafood.[29] This study implies that consumer concerns with safety have caused a decrease in demand for Gulf seafood and seafood in general.[30]

Supply effects on prices depend on the scale of the reduction in supply relative to the total market supply. The decrease in supply and related price increases are likely to have been greater for fresh markets near the area of the spill and for local specialties such as oysters.[31] With regard to demand, concerns with seafood safety are likely to put negative pressure on prices. Initially, prices of some Gulf seafood products increased because supplies were constrained by fishery closures. On average, reported Gulf shrimp prices have remained higher in 2010 than during 2009.[32] Oysters have also increased in price, likely in part because of the decrease in Louisiana production which is over 37% of national production and 62% of production in the Gulf region.[33] As areas have been reopened and supply has rebounded, there have been some reports that prices of some products have decreased relative to previous years. In addition to the effects related to the oil spill, prices also depend on additional factors such as the availability of substitutes, consumer income, and consumer tastes and preferences.[34] The length of time it will take to regain markets and consumer trust remains an open question.

Environmental Impacts of the Oil Spill

The Deepwater Horizon oil spill has harmed living organisms that inhabit ocean and coastal areas of the Gulf of Mexico, although the magnitude of damages are subject to considerable uncertainty. Coastal areas are especially vulnerable because oil can be stranded in wetlands and other coastal ecosystems after being washed in by waves and tides. The uptake of dissolved components of oil is toxic for fish, shellfish, other invertebrates, and plankton. Oil also may coat small animals and plants that inhabit shoreline areas and suffocate them. Many scientists are concerned that oil suspended in the water column may have caused mortality of plankton, including eggs and larvae of many fish such as bluefin tuna. Oil also may have remained on or near the bottom of the Gulf and affected deep corals and other bottom-dwelling organisms. Sublethal effects reduce the overall health of organisms, resulting in decreased growth and reproduction. Initial harm to marine organisms, such as direct mortality and reduced health, decrease the reproductive capacity of marine populations and may reduce future abundance.

Early life stages of many species develop in coastal areas such as estuaries and wetlands and move offshore as they grow to their adult stage. As a result, the health of coastal areas can have long-term implications for the fishery industry. Ninety-seven percent of commercial fish and shellfish landings by volume are composed of species that depend on estuaries and wetlands at some point in the life cycle, and landings from the Louisiana coastal zone account for nearly one-third of the fish volume harvested in the continental United States.[35] The presence of oil in the environment may alter migration patterns, decrease food availability, and disrupt life cycles. According to a study of oil impacts on Louisiana fisheries, populations of shrimp, crab, and menhaden are most likely to be harmed by the effects of oil on their eggs and larvae.[36]

The impacts of oil in coastal areas affects more than individual fish species. The Gulf is composed of inter-related ecosystems that stretch from estuaries and coastal wetlands to the pelagic zone (open ocean). Species directly affected by the spill could affect other species because of ecological interactions. For example, oil in coastal areas has affected structure-forming organisms, such as marsh grasses and oysters. These organisms provide shelter and a surface for attachment by other marine organisms. Survey teams have documented 1,053 total linear miles of oiled shoreline in the Gulf.[37] Robert Barham, Louisiana Wildlife and Fisheries Secretary, has reportedly voiced concerns over long-term implications of the oil spill on Gulf wildlife, such as effects on the food web.[38]

Fisheries surveys from Dauphin Island Sea Lab, conducted after the oil spill off the coasts of Mississippi and Alabama, have found that the abundance of some fish species appear to have increased in 2010. Some scientists have speculated that these populations have increased because of the oil spill-related fisheries closures and reduced harvest.[39] Some scientists expressed concern that the effects of the fishing closures may make it difficult to determine the direct impact of the oil spill on marine populations. The complexity of coastal ecosystems and scale of the Gulf oil spill are likely to contribute to considerable uncertainty regarding the magnitude and duration of spill-related damages to living resources.

Financial Assistance

Several mechanisms exist to provide short- and long-term assistance to the fishing industry. Financial assistance to compensate for economic injuries to individuals and businesses affected by an oil spill are defined in statute by the Oil Pollution Act. Another source of assistance can be provided in cases where a fishery failure is declared by the Secretary of Commerce under the Magnuson-Stevens Fishery Conservation and Management Act (MSFCMA, P.L. 109-479). Additional sources of assistance have been provided through BP grants, the Vessels of Opportunity Program, and the Small Business Administration.

Claims and Compensation

Following the Exxon Valdez oil spill in 1989, a compensation and claims process was established under OPA for costs/damages resulting from oil spills, including lost profits and earnings resulting from property loss or natural resource injury.[40] In general, claims for damages must be presented first to the responsible party (e.g., BP).[41] If the party to whom the claim is presented denies all liability, or if the claim is not settled by payment within 90 days after the claim was presented, the claimant may elect either to initiate an action in court against the responsible party or to present the claim directly to the Oil Spill Liability Trust Fund.[42]

In response to economic harm caused by the oil spill and to fulfill obligations as a responsible party, BP established a claims process and multiple claims centers.[43] On May 3, 2010, BP began paying emergency compensation to individuals and businesses. BP stated that emergency payments would continue as long as individuals and businesses could show they were unable to earn a living because of injury to natural resources caused by the oil spill. Payments were based on one month of income and would be adjusted with additional documentation.[44] On June 16, 2010, President Obama announced that BP had agreed to set aside $20 billion to pay economic damage claims caused by the oil spill. Some members of the fishing industry criticized BP's administration of the claims process because they maintained that the claims process was slow, some individual payments were inadequate, and required proof of past earnings might not reflect potential earnings (if the spill had not occurred) for 2010.

On August 23, 2010, the Gulf Coast Claims Facility (GCCF) took over the administration of claims from BP to address the issues with BP's claims process.[45] Up until the change, BP had paid out $395.6 million, of which approximately $111 million went to the fisheries industry. Although still funded by BP, GCCF was established by the Administration and BP to provide an independent claims process.[46] Some have argued that the process is not independent because the GCCF is directly financed by BP. GCCF provides information on how to file a claim, how to check the current status of a claim, and other general information concerning the process. GCCF offered emergency payments equivalent to six months of lost income to claimants until November 23, 2010. As of February 12, 2011, the GCCF had made emergency payments of approximately $751 million to individuals and businesses in the fishing industry.[47] Some claimants have voiced concerns with the transparency of the claims process, the lack of information in GCCF responses to claims, and the adequacy of payments, but the actual proportion of unsatisfied fishing industry claimants is not known at this time.

Under GCCF procedures, claimants will have three years to estimate damages and submit claims for final payment. Acceptance of a final claim would resolve all claims by that party against BP including past and future alleged damages.[48] Individuals and businesses that have received emergency payments from the GCCF are eligible for a quick payment final claim, which offers a fixed amount of $5,000 for individuals and $25,000 for businesses. Those who do not choose or are not eligible for the quick payment may submit a full review final payment claim for all documented losses and damages. The alternative to a final payment is to make an interim payment claim for past damages that have not been compensated. Individuals and businesses receiving interim payments are not required to sign a release of liability and may file a final claim at a later date. For many in the fishing industry, their dilemma is related to uncertainty in determining the extent and duration of damages to fisheries resources. Some have questioned whether three years is a sufficient period to fully determine the damages of the oil spill. Many individuals and businesses are faced with the decision on whether to take a final settlement or, as many have already done, to file a lawsuit, which may present uncertainty as to payment size and timing.

On February 2, 2011, the GCCF released a draft proposal, *Payment Options, Eligibility and Substantiation Criteria, and Final Payment Methodology*, for public comment.[49] The proposal establishes principles for governing final and interim claims for individuals and businesses. GCCF also released a report, *An expert opinion of when the Gulf of Mexico will return to pre-spill harvest status following the BP Deepwater Horizon HC 252 oil spill*, to provide supplemental information for the GCCF final payment methodology.[50] The report predicts that regional catches for major Gulf fisheries will likely continue along the same harvest trends of recent years by 2011.[51] However, the report recognized there may be exceptions for specific areas and fisheries, especially for some oyster beds that may not recover for 6 to 10 years. Furthermore, the report acknowledges that a definitive assessment of recovery time is impossible, and the true loss to ecosystems and fisheries may not be known for years or even decades. Some have criticized the report because it assesses recovery of Gulf fisheries without allegedly understanding the ecological and long-term affects of the oil spill.[52]

Vessels of Opportunity Program

The BP Vessels of Opportunity Program was designed to provide local boat operators with the opportunity to assist with response activities such as transporting supplies, assisting with wildlife rescue, and deploying containment booms. Only captains and employees who completed training and met other conditions were allowed to participate in the program. Approximately 3,500 commercial and charter fishing boats were employed over the life of the program. As of January 20, 2011, the program had made payments of $594 million for vessels and crew in Louisiana, Mississippi, Alabama, and Florida. In September 2010, the program was concluded in Florida, Alabama, and Mississippi, but as of January 2011, a small number of vessels remained active in Louisiana. The GCCF has decided that the earnings from the Vessels of Opportunity Program would not be deducted from payments made to claimants.[53]

Other BP Assistance for the Gulf Fishing Industry

On November 1, 2010, BP agreed to provide Louisiana with $48 million for seafood safety (testing programs) and marketing.[54] Testing for oil and dispersants in seafood will be funded at $6 million per year for three years and Louisiana seafood marketing will be funded at $10 million per year for three years. The three-year commitment would reset for both programs if oil triggers the closure of new fishing areas. Louisiana had proposed funding of $173 million for a long-term seafood testing and marketing campaign, but BP refused to fund the program. Previously, BP agreed to provide $20 million over three years to fund seafood inspections and marketing efforts in Florida.[55] BP also has agreed to provide $13 million to Louisiana to fund a three year study to monitor effects of the oil spill on Gulf fisheries.[56]

NOAA Fishery Disaster Assistance and Other Federal Efforts

On May 24, 2010, Secretary of Commerce Locke determined that the ongoing oil spill had caused a fishery failure in the states of Louisiana, Mississippi, and Alabama. On June 2, 2010, the Secretary added Florida to the earlier determination. Both determinations were made under Section 312(a) of the Magnuson-Stevens Fishery Conservation and Management Act (MSFCMA). Under this section of the act, states must provide a 25% match to federal funds. The Administration requested $15 million to address the fishery failure and $5 million in economic development assistance through the Economic Development Administration. The Supplemental Appropriations Act of 2010 (P.L. 111-212) included a total of $28 million for fishery disaster assistance and $5 million for economic development assistance. NOAA is expected to allocate $15 million of the fishery disaster assistance for a strategic marketing plan and health and safety assurance program for Gulf coast seafood. The remaining $13 million of fishery disaster assistance is intended to address economic impacts on fishermen and fishery-dependent businesses only if resources provided under other authorities are insufficient. The Supplemental Appropriations Act of 2010 also included $10 million for expanded stock assessments for Gulf fish species and $1 million for a National Academy of Sciences study of the long-term effects of the oil spill.[57] In addition, the Small Business Administration has offered economic injury disaster loans and has deferred required payments for some existing loans.[58]

RESTORATION

Gulf fisheries production is dependent on healthy Gulf ecosystems, including the quantity and quality of fisheries habitat. Restoration of Gulf ecosystems would likely maintain and enhance current fisheries production.[59] The two main restoration efforts focusing on oil spill damages are the Natural Resource Damage Assessment (NRDA) and the Administration's Gulf Coast Restoration Plan.

Natural Resource Damage Assessment

The federal government's role in restoration is defined in statute by OPA and in NOAA regulations for developing a NRDA. NRDA addresses natural resource damages, restoration of resources that are injured, and lost services that result from an oil spill. The parties responsible for causing the oil spill are responsible for NRDA damages.[60] In contrast to financial compensation for individuals and businesses, NRDA focuses on restoration and compensation for harm to public natural resources. Designated federal, state, tribal, and sometimes foreign trust agencies are responsible to act on behalf of the public. OPA directs trustees to undertake two main actions: (1) return injured natural resources to their baseline condition (the condition that existed prior to the spill), and (2) compensate for interim losses. Restoration actions focus on returning natural resources to the baseline level with as much certainty and as quickly as possible. Compensation covers actions to address interim losses of natural resources and services until resources have recovered. Compensatory actions provide services of the same type and quality and of comparable value as those lost or injured. Damage assessment is required to quantify the extent of injuries to natural resources and to determine the type and amount of restoration and compensatory actions needed. The process of recovery is broken down by the regulations into three main phases:[61]

- **Pre-assessment phase**—determines whether natural resource injuries have occurred or are expected and whether to continue to the next phase.
- **Restoration planning phase**—evaluates potential injuries to natural resources. This phase includes an assessment of the nature and extent of natural resource injuries and development of plans for restoring resources and compensating the public for interim losses.
- **Restoration phase**—the final restoration plan is presented to responsible parties to implement or fund the plan. This provides the opportunity for settlement of damage claims without litigation. However, OPA authorizes trustees to bring civil action for damages.

As of January 2011, trustees were continuing with the restoration planning phase of the process that was begun in August of 2010. Efforts required during this phase include defining the baseline condition of the ecosystem and damages caused by the oil spill. Quantifying these conditions and damages are often hindered by limited scientific understanding of physical and biological processes in coastal and marine areas, natural variability of marine systems, and a paucity of related scientific data. These factors are coupled with uncertainties about acute and chronic effects of oil on marine organisms. In the face of these uncertainties, it is possible that questions related to restoration and compensation will arise, including basic questions about what constitutes ecosystem recovery and how to determine when it has occurred.

Gulf Coast Restoration Plan

On June 15, 2010, the Administration committed to developing a long-term Gulf of Mexico restoration plan for post-spill recovery needs as well as long-term restoration. In contrast to the environmental damages addressed by NRDA, the Administration's plan would address a broader array of restoration needs, many of which pre-date the oil spill. In September 2010, the Administration released *America's Gulf Coast; A Long Term Recovery Plan After the Deepwater Horizon Oil Spill.*[62] The report put forward a plan to restore the environment, economy, and public health of residents. It stressed the need for inclusive engagement and collaboration among governmental, private, and non-profit organizations and for a significant sustained commitment of resources over many years. The report recommends that Congress dedicate a significant, amount of any civil penalties recovered from responsible parties under the Clean Water Act to restore the Gulf coast.[63] Further, it recommends that Congress create a Gulf Coast Recovery Council to coordinate efforts taken by concerned parties. The report recommended that the Administration immediately establish a Gulf Coast Ecosystem Task Force to coordinate Gulf restoration efforts until the Council is established. In January 2011, the National Commission on the BP Deepwater Horizon Oil Spill and Offshore Drilling released its report to the President.[64] Two of the Commission's recommendations for restoration focus on these points.

> Congress should dedicate 80 percent of the Clean Water Act penalties to long-term restoration of the Gulf of Mexico.
> Congress and federal and state agencies should build the organizational, financial, scientific, and public outreach capacities needed to put the restoration effort on a strong footing.

On October 5, 2010, President Obama signed Executive Order 13554 to establish the Gulf Coast Ecosystem Restoration Task Force.[65] The task force is chaired by the Administrator or representative of the Environmental Protection Agency and composed of representatives of the Departments of Defense, Justice, Interior, Agriculture, Commerce, and Transportation; and the Office of Management and Budget; the Council on Environmental Quality; the Office of Science and Technology Policy; the Domestic Policy Council; representatives appointed by the Governors of the Gulf states of Texas, Louisiana, Mississippi, Alabama, and Florida; and may include representatives of affected tribes. The task force's main goal is to develop a "Gulf of Mexico Regional Ecosystem Strategy." The strategy will set goals, develop performance indicators, and set up a process to coordinate intergovernmental restoration efforts. The Task Force first met on November 8, 2010.

CONGRESSIONAL ACTIONS AND POTENTIAL CONCERNS

The 111th Congress held 62 hearings related to the Gulf oil spill and at least 14 were directly related to natural resources and impacts on small businesses. The 112th Congress may continue to conduct oversight of efforts on fishing industry recovery, adequate compensation, and Gulf restoration. Ongoing efforts by federal agencies and states to ensure seafood safety and to regain and maintain the reputation of Gulf seafood are the most immediate challenges. As the NRDA process moves from the planning to restoration phase, questions may arise regarding the level of the potential settlement and the types of restoration activities identified by the trustees.

In contrast to restoration efforts developed under NRDA that are based on existing law, funding and governance of the Administration's comprehensive and long-term Gulf of Mexico restoration plan would require congressional action. To fully implement and fund the Administration's plan, Congress may consider whether to commit Clean Water Act civil penalties for this purpose as was recommended by the Administration's Gulf recovery plan and the National Commission on the BP Deepwater Horizon Oil Spill and Offshore Drilling. Congress also may consider how the fund should be allocated and whether to establish a Gulf Restoration Council to coordinate restoration efforts. Several bills were introduced during the 111th Congress that included these elements of the Administration's plan, but no action was taken on any of them. An ongoing issue, especially for Gulf states, is the allocation and permitted uses of funding under the Administration's restoration plan.

Three bills have been introduced in the 112th Congress that address elements of the Administration's restoration plan, including the Gulf Coast

Restoration Act (H.R. 56) and Title IV of Recommendations of the BP Oil Spill Commission Act of 2011 (H.R. 501), which are identical, and the Gulf of Mexico Economic and Environmental Restoration Act of 2011 (H.R. 480). All three bills would establish a Gulf Coast Ecosystem Restoration Fund and require 80% of any amounts collected by the United States as penalties, settlements, or fines under the Federal Water Pollution Control Act (33 U.S.C. § 1319, § 1321) to be deposited into the fund. They would also establish a governing body to distribute funding and coordinate restoration efforts such as a Gulf Coast Ecosystem Restoration Task Force (H.R. 56 and H.R. 501) or a Gulf of Mexico Recovery Council (H.R. 480). In contrast to H.R. 56 and H.R. 501, H.R. 480 would provide a formula for allocating funding among Gulf states, and establish and fund four Gulf of Mexico programs, including an observation system, a grant program for Sea Grant colleges, a seafood marketing program, and a clean energy program.

End Notes

[1] U.S. Department of the Interior, Press Release, "U.S. Scientific Teams Refine Estimates of Oil Flow from BP's Well Prior to Capping," August 2, 2010, at *http://app.restorethegulf.gov/* release/2010/08/02/us-scientific-teams-refine-estimates-oil-flow-bps-well-prior-capping.

[2] As of February 15, 1,041 square miles of federal waters remained closed to fishing, http://sero.nmfs.noaa.gov/ ClosureSizeandPercentCoverage.htm.

[3] Gulf Coast Claims Facility, *Measures of the Effects of the Gulf Oil Spill on Individuals and Businesses and Proposed Compensation Schema*, January 24, 2011, http://www.gulfcoast claimsfacility.com/exhibit_c.pdf.

[4] *NRDA by the Numbers - January 2011*, *http://www.gulfspillrestoration.noaa.gov/wp-content/* uploads/2011/02/ NRDA_by_the_Numbers_1_11_FINAL.pdf.

[5] According to 26 U.S.C. 9509(b)(8) the potential Clean Water Act penalties would be deposited into the Oil Spill Liability Trust Fund.

[6] The term fishery includes the resource, fishermen, and related businesses.

[7] National Marine Fisheries Service, *Fisheries of the United States 2009*, Silver Spring, MD, July 2010, p.6, http://www.st.nmfs.noaa.gov/st1/fus/fus09/index.htmlhttp:// www.st.nmfs.noaa. gov/st1//fus/fus09/03_recreational2009.pdf.

[8] National Marine Fisheries Service, *Fisheries Economics of the United States 2008*, Economics and Sociocultural Status and Status Trends Series, Silver Spring, MD, April 2010, p. 119, http://www.st.nmfs.noaa.gov/st5/publication/ fisheries_economics_2008.html. Hereinafter cited as NMFS 2010. Job and income figures include commercial fisheries for the entire state Florida.

[9] NOAA Fisheries Office of Science & Technology, Fisheries Statistics, Queried commercial landings and revenue, February 10, 2010, http://www.st.nmfs.noaa.gov/st1/commercial/ landings/annual_landings.html.

[10] National Marine Fisheries Service, *Fisheries of the United States 2008*, Silver Spring, MD, July 2009, p. 38 (data on recreational trips and fish caught do not include Texas), http://www.st.nmfs.noaa.gov/st1/fus/fus09/index.htmlhttp://www.st.nmfs.noaa.gov/st1//fus/ fus09/03_recreational2009.pdf.

[11] Includes vehicles, boats, fishing tackle, and other items.

[12] NMFS 2010, p. 120.

[13] See http://sero.nmfs.noaa.gov/ClosureSizeandPercentCoverage.htm.

[14] Gulf Coast Claims Facility, *Measures of the Effects of the Gulf Oil Spill on Individuals and Businesses and Proposed Compensation Schema*, January 24, 2011, http://www.gulfcoastclaimsfacility.com/exhibit_c.pdf.

[15] "On November 23, 2010, 4,213 square miles were closed to the royal red shrimp fishery as a precautionary measure because tarballs were caught in nets by a commercial vessel trawling for royal red shrimp. On February 1, 2011, the area was reopened to royal red shrimp fishing after shrimp sampling and testing from the area showed no oil or dispersant contamination," http://sero.nmfs.noaa.gov/sf/deepwater_horizon/FB11-010_Reopening_RoyalRed_020211.pdf.

[16] Food and Drug Administration, Overview of Testing Protocol to Re-open Harvest Waters that Were Closed in Response to the Deepwater Horizon Oil Spill, September 2010, http://www.fda.gov/Food/ucm217598.htm.

[17] Food and Drug Administration, *Gulf of Mexico Oil Spill: Questions and Answers*, http://www.fda.gov/Food/FoodSafety/Product-SpecificInformation/Seafood/ucm221563.htm.

[18] Bob Marshall, "Safety of Gulf Seafood Debated 8 Months After BP Oil Spill," *The Times-Picayune*, December 19, 2010, http://www.nola.com/news/gulf-oil-spill/index.ssf/2010/12/safety_of_gulf_seafood.html.

[19] U.S. Congress, House Committee on Energy and Commerce, Subcommittee on Energy and Environment, *The BP Oil Spill: Accounting for the Spilled Oil and Ensuring the Safety of Seafood from the Gulf*, 111th Cong., 2nd sess., August 19, 2010.

[20] National Oceanic and Atmospheric Administration, "NOAA Reopens More Than 8,000 square miles of the Gulf of Mexico to Fishing," press release, November 15, 2010, http://www.noaanews.noaa.gov/stories2010/ 20101115_reopening.html.

[21] FDA, Gulf of Mexico Oil Spill Update, Information current as of September 8, 2010, http://www.fda.gov/food/ ucm210970.htm.

[22] NOAA Fisheries Office of Science and Technology, *Fishery Market News*, Monthly Gulf Coast Shrimp Statistics, December 2010, http://www.st.nmfs.noaa.gov/st1/market_news/doc45.txt. (Hereinafter cited as NOAA Fishery Market News).

[23] See http://www.houmatoday.com/article/20101216/articles/101219646.

[24] 21 U.S.C. § 331 and 21 U.S.C. § 342(a).

[25] Crude oil also degrades, but it is made-up of different components, which degrade at different rates.

[26] Susan D. Shaw, *Consensus Statement*, Maine Environmental Research Institute, "Scientists oppose the use of dispersant chemicals in the Gulf of Mexico," July 16, 2010, http://meriresearch.org/Portals/0/Documents/CONSENSUS%20STATEMENT%20ON%20DISPERSANTS%20IN%20THE%20GULF%20updated%20July%2017 .pdf.

[27] National Research Council, *Oil Spill Dispersants*, Efficacy and Effects, Washington, DC, 2005, p. 11.

[28] Kiros Hailemariam, *Oil Spill Chemical Dispersants*, Department of Health and Human Services, Food and Drug Administration, May 14, 2010.

[29] Kevin McGill, "Survey measures post-spill seafood attitudes," *Bloomberg Businessweek*, January 31, 2011, http://www.businessweek.com/ap/financialnews/D9L3IP0O0.htm.

[30] A change in demand results in a shift in the demand curve, in this case to the left, as opposed to a change in quantity demanded, movement along the demand curve.

[31] Louisiana is also a major producer of oysters nationally.

[32] NOAA Fishery Market News 2010.

[33] NOAA Fisheries Office of Science & Technology, Fisheries Statistics, Queried commercial landings and revenue February 10, 2010, http://www.st.nmfs.noaa.gov/st1/commercial/landings/annual_landings.html.

[34] Shrimp prices have generally been higher for reasons unrelated to the oil spill.

[35] National Oceanic and Atmospheric Administration, *Shorelines and Coastal Habitats in the Gulf of Mexico*, Factsheet, http://www.noaa.gov/factsheets/new%20version/shorelines_coastalhabitat.pdf.

[36] Innovative Emergency Management Inc., *A Study of the Economic Impact of the Deepwater Horizon Oil Spill*, Greater New Orleans Inc., October 15, 2010.

[37] *NRDA by the Numbers - January 2011*, http://www.gulfspillrestoration.noaa.gov/wp-content/uploads/2011/02/ NRDA_by_the_Numbers_1_11_FINAL.pdf.

[38] Nikki Buskey, "Official Says That BP Won't Pay for Seafood Program," *Daily Comet*, October 27, 2010.

[39] See http://www.fishbio.com/environmental-consulting-and-environmental-research-news/marine-news/10043-fish-numbers-triple-after-oil-spill-fishing-closures.html.

[40] OPA § 1013 (33 U.S.C. 2713). Implementing regulations are found in 33 C.F.R. Part 136.

[41] 33 U.S.C. § 2713(a). Under OPA, the term "claim" means "a request, made in writing for a sum certain, for compensation for damages or removal costs resulting from an [oil spill] incident." 33 U.S.C. § 2701(3).

[42] The Oil Spill Liability Trust Fund (OSLTF) was established under OPA and consolidated the liability and compensation requirements of certain other prior oil pollution laws and their supporting funds. The OSLTF may be used for a variety of purposes related to oil spills, including payment of claims for uncompensated removal costs and damages. See http://www.uscg.mil/npfc/About_NPFC/osltf.asp.

[43] Pursuant to OPA (33U.S.C. 2713-2715).

[44] U.S. Congress, House Committee on the Judiciary, *Testimony of Darryl White, Vice President, Resources, BP America*, 111th Cong., 2nd sess., May 27, 2010. Hereinafter cited as, BP 2010.

[45] See http://www.whitehouse.gov/the-press-office/fact-sheet-claims-and-escrow.

[46] See http://www.gulfcoastclaimsfacility.com/faq#Q1.

[47] See http://www.gulfcoastclaimsfacility.com/reports.

[48] See http://www.gulfcoastclaimsfacility.com/.

[49] Gulf Coast Claims Facility, *Payment Options, Eligibility and Substantiation Criteria, and Final Payment Methodology*, February 2, 2010, http://www.gulfcoastclaimsfacility.com/METHODOLOGY.%20FEB.2.%20FINAL%20DRAFT.pdf.

[50] John W. Tunnell, Jr., An expert opinion of when the Gulf of Mexico will return to pre-spill harvest status following the BP Deepwater Horizon MC 252 oil spill, Harte Research Institute for Gulf of Mexico Studies, Commissioned by Kenneth R. Feinberg, Claims Administrator, Gulf Coast Claims Facility, Corpus Christi, TX, January 31, 2011, http://www.gulfcoastclaimsfacility.com/exhibit_c.pdf.

[51] The four major fisheries considered in the report are shrimp, oysters, blue crab, and finfish. The finfish category grouped together commercial fish species.

[52] Brett Michael Dykes, "Feinberg: The Gulf will fully recover from the oil spill by 2012," *The Lookout*, February 2, 2011, *http://news.yahoo.com/s/yblog_thelookout/20110202/us_yblog_* thelookout/feinberg-the-gulf-will-fully-recover-from-oil-spill-by-2012.

[53] See http://www.sbpg.net/index.php?option=com_content&view=article&id=1966:gulf-coast-claims-facility-will-not-deduct-vessel-of-opportunity-earnings-from-claims-payments&catid=84:oil-spill&Itemid=252.

[54] BP, "Seafood Safety, Tourism, and Restoration Funding Announced," press release, November 1, 2010, http://www.louisianagulfresponse.com/go/doc/3047/940587/.

[55] Cristine Blank, "BP to Fund Florida Seafood Testing Marketing," *Seafood Source.com*, October 25, 2010, p. *http://www.seafoodsource.com/newsarticledetail.aspx?id=4294999754*.

[56] Louisiana Department of Fisheries and Wildlife, "Governor Jindal Announces Funding for Fisheries Monitoring Program, Calls on BP to Fund Long-term Seafood Program," press release, August 18, 2010, *http://www.wlf.louisiana.gov/category/newsletter/ldwf-news?page =11*.

[57] See http://www.commerce.gov/print/news/press-releases/2010/09/17/us-commerce-secretary-gary-locke-announces-26-million-gulf-states-fis.

[58] For more information related to disaster assistance see CRS Report RL34146, *FEMA's Disaster Declaration Process: A Primer*, by Francis X. McCarthy.

[59] An exception might include Gulf oysters because oyster reefs depend on the current circulation and mixing of fresh and salt water. Alteration of freshwater flows would likely change the quality of oyster habitat.

[60] Thus far responsible parties include British Petroleum Exploration and Production Inc., Transocean Holdings Inc., Triton Asset Leasing GmbH, Transocean Offshore Deepwater Drilling Inc., Transocean Deepwater Inc., Anadarko Petroleum, Anadarko E&P Company LP, and MOEX Offshore 2007 LLC.

[61] Department of Commerce, "Natural Resource Damage Assessments; Final Rule," 61 *Fed. Reg.* 441-442, January 5, 1996.

[62] Ray Mabus, *America's Gulf Coast, A Long-term Recovery Plan After the Deepwater Horizon Oil Spill*, September 2010, http://www.restorethegulf.gov/sites/default/files/documents/pdf/gulf-recovery-sep-2010.pdf.

[63] Normally fines and penalties are deposited in the federal treasury.

[64] National Commission on the BP Deepwater Horizon Oil Spill and Offshore Drilling, *Deep Water, The Gulf Oil Disaster and the Future of Offshore Drilling*, Report to the President, January 2011, https://s3.amazonaws.com/ pdf_final/DEEPWATER_ReporttothePresident_FINAL.pdf.

[65] Executive Order 13554, "Establishing the Gulf Coast Ecosystem Task Force," 75 *Federal Register* 62313 - 62317, October 8, 2010.

In: Impacts of the Gulf Oil Spill on Fishing ... ISBN: 978-1-61470-971-8
Editors: C. Scarnati and E. Popeo

Chapter 2

THE DEEPWATER HORIZON OIL SPILL: COASTAL WETLAND AND WILDLIFE IMPACTS AND RESPONSE*

M. Lynne Corn and Claudia Copeland

SUMMARY

The explosion of the Deepwater Horizon drilling rig in the Gulf of Mexico on April 20, 2010, and the resulting oil spill began a cascade of effects on the coastal areas of the Gulf and on the wealth of species that inhabit those areas, effects that continue even though oil stopped flowing in July and the well was sealed in September. Gulf coast wetlands, like those elsewhere, have value for water quality, flood control, shoreline protection, and recreation. They serve as nurseries for many species, including fish and shellfish of commercial significance, waterfowl, and a host of resident and migratory species. They also have cultural importance to the people of the Gulf. The effects of the spill come on top of historic wetland losses due to subsidence, drainage, and saltwater intrusion, along with rising sea levels, coastal erosion, and global climate change.

Impacts of oil spills on wetland ecosystems depend on multiple factors, including the type of oil, exposure of the oil to weathering factors before it

* This is an edited, reformatted and augmented version of Congressional Research Service Report R41311, dated October 15, 2010.

reaches the shore, the season in which the spill occurs, etc. Mitigation and cleanup of damage to wetlands is far from an exact science and involves many tradeoffs; there is no single, best solution. A number of options exist, ranging from mechanical recovery and use of dispersants to doing nothing. Estimating wildlife impacts is particularly difficult in this case because the spill occurred far offshore, and the initial wildlife mortality came far out in the Gulf, where animals sank without reaching the shore. When oil arrived at the shore, more animals began to be counted. Moreover, because the Gulf wetlands host many species of birds during seasonal migrations, impacts of the spill could be felt in areas well away from the Gulf.

Among other issues is a seemingly simple question: who decides what to do when such an event occurs? But the answer is complex. The organizational structure for deciding how to respond to oil spills is specified in the National Contingency Plan (NCP), which was created administratively and has been broadened by the Clean Water Act, the Superfund law, and the Oil Pollution Act. Under the NCP structure, the Coast Guard is the lead federal agency for overseeing response and cleanup. Until oil from the well stopped flowing, very little cleanup of wetlands was occurring, because of both the ongoing risk of greater harm from cleanup and the potential for re-oiling. As cleanup proceeds, a number of questions arise. To cite only two, what factors will determine cleanup strategies, and how are needs to improve scientific understanding of the impacts of oil spills being considered?

Decisions about cleanup of wildlife are no easier. Cleanup of individual animals is labor-intensive, and some scientists argue that the survival of an animal that has been cleaned is so uncertain as to call into question whether treatment is, in fact, humane. Rescue groups are dedicated to salvaging those that can still be saved. The effects on a species as a whole vary markedly from one species to another, depending on that species' abundance and ecological needs; appropriate responses at the species level are unclear.

History, particularly from the relatively well-studied *Exxon Valdez* spill of 1989, offers insight into the future of Gulf resources as well. First, some cleanup efforts might do more harm than good in the long run. Second, it is not possible to predict all of the ramifications for the complex Gulf ecosystem in the decades to come, but history suggests that at least some effects will continue for decades. Finally, litigation could play a major role in disseminating—or not disseminating—scientific information about the spill and its effects.

INTRODUCTION

Since the April 20, 2010, explosion on the mobile offshore drilling unit Deepwater Horizon in the Gulf of Mexico, public and private efforts have focused on containing, capturing, cutting off, and capping the oil. These collective efforts under water and on the surface of the Gulf are intended in large part to keep the oil from reaching coastal marshes and wetlands—what Coast Guard Admiral Thad Allen, the National Incident Commander for responding to the spill, termed the "worst case scenario for us to deal with."[1] Nevertheless, within a few weeks of the explosion there were reports that oil had reached coastal areas and had begun to affect land and wildlife.[2]

Uncontrolled leaking from the MC252 Macondo well site continued for three months after the explosion. In mid-July, the flow of oil from the well was stopped with a temporary cap on the gushing wellhead, but oiling of the Gulf's shorelines continued even after oil was no longer flowing from the well. On September 17, more than 35% of Gulf Coast shoreline miles were oiled to some degree, but less than 7% was oiled to a moderate or heavy extent, according to the federal government.[3] On September 19, the government reported that the ruptured well was "effectively dead" and posed no continuing threat to the Gulf because of the completion of a relief well that successfully sealed off the last path for oil to escape.[4] Despite that good news, scientists and public officials recognize that damage to marine and wildlife habitats has occurred (although immediate impacts were less severe than anticipated), not all of the 4.9 million barrels of oil estimated to have been released from the well is gone from the Gulf, and concern remains about lingering effects that may occur for some time to come.

The risk posed by the wellhead has ended, as has the emergency phase of response. But the oil spill response is ongoing and day-to-day cleanup operations continue—cleaning shoreline, sampling Gulf waters, and responding to any recoverable oil that is found.[5] The extent of oiled vegetation on shore, resulting damages, and how quickly plants will regenerate are largely unknown. Biological effects on wildlife won't be known for at least a full breeding/spawning season. Additional impacts may occur as a result of oil that remains in the Gulf—oil that has not been recovered, burned, skimmed, evaporated, or degraded naturally. Beyond such generalizations, the history of prior spills does show that damages are likely to be unknown or poorly understood for some time, probably years. And whether science will be available to answer important questions is uncertain (see discussion of "Science and Litigation," below).

This report addresses the importance of wetlands in general, the ecology of the coastal wetlands in the Gulf, impacts of oil spills on wetland habitats, response options, the implications of hurricane season for the spill's impacts, and cleanup and recovery issues.[6] The emphasis is on the nearshore environment, although a few species found in deeper waters will be mentioned. In addition, some lessons from past spills such as the *Exxon Valdez* in Alaska will be discussed, along with issues that may arise as response and recovery transition to restoration of the Gulf.

WHY ARE WETLANDS IMPORTANT?

Wetlands are sometimes referred to as "in-between" areas that consist of both dry uplands and open water environments—they are transition zones that are neither "land" nor "water." More commonly, wetlands are known as marshes, bogs, and swamps. Because they have both upland and aquatic characteristics, they often have a richer flora and fauna than other environments. In practice, wetlands may be hard to define, but scientists generally agree that the presence of a wetland can be determined by a combination of certain distinctive soils, plants, and hydrology. In different regions, some wetland areas may be continuously inundated by water, while other areas may not be flooded at all, or only at certain times of the year. In coastal areas, flooding may occur daily as tides rise and fall.

Functional values, both ecological and economic, at each wetland depend on its location, size, and relationship to adjacent land and water areas. Many of these values have been recognized only recently. Historically, many federal programs encouraged wetlands to be drained or altered, because they were seen as having little intrinsic value. Today, scientists and policymakers recognize many valuable functions, including flood control and shoreline protection, as well as nurseries and protection for fish, shellfish, and wildlife, as described in the box below.

FUNCTIONS AND VALUES OF WETLANDS

Water Quality. Wetlands are efficient water filters, because wetland plants and soils clean fresh water of contaminants before it moves into surface waters or groundwater. Wetlands filter runoff and remove sediment, nutrients, pesticides, metals, and other pollutants.

Flood Buffers and Erosion Control. Wetlands reduce the severity of floods by acting as natural retention areas; consequently, destruction or loss of wetlands makes downstream flooding worse. Similarly, nearshore wetlands act as buffers to reduce shoreline erosion and stabilize land.

Shoreline Protection. Coastal wetlands are particularly important in the Gulf region, because they can help reduce coastal flooding by absorbing the energy, wind, and water from incoming hurricane storm surges. Many of the coastal areas in the Gulf are in flat or low-lying areas, which are particularly vulnerable to tropical storms and hurricanes. It is estimated that for every 3.4 miles of healthy coastal wetlands that a given storm surge must travel over, the height of the surge is diminished by one foot.

Nurseries for Fish and Shellfish. Coastal wetlands serve as important nurseries for numerous fish and shellfish, as many species depend on wetlands for some or all of their life cycle, where they feed, take refuge, or reproduce. Approximately 75% of the nation's commercial fish and shellfish depend on estuaries at some stage in their life cycle, and estuaries depend on the adjacent wetlands to maintain water quality and provide the basis for food chains. About 97% of the commercial fishery landings from the Gulf are species that depend on estuaries for reproduction, nursery areas, food production, migrations, or shelter. Commercial fishing in the Gulf coastal region contributed more than $680 million to the nation's economy in 2007.

Wildlife Habitat. Coastal wetlands are home to many different animals and, importantly, are stopover or winter habitat for many migrating birds. In the Gulf region, birds that live in wetlands include many species of egrets, shorebirds, ibises, pelicans, herons, and others. Gulf wetlands also are home to many threatened or endangered species such as piping plovers, and several species of sea turtles and sturgeon. Marine mammals are also found in wetlands and along the Gulf coast; species include manatees, three species of whales, and seals. Some of the marine mammals are also protected as endangered or threatened. (See "Coastal Habitat as Home," below.)

Recreation. Many recreational activities depend on wetlands, including fishing, hunting, bird-watching, boating, and wildlife photography. In 2004, tourism and recreation employment in the five Gulf coast states totaled more than 416,000 jobs.

Cultural Importance. Many cultures, such as the Cajun culture in the Gulf, have been heavily influenced by or rely upon wetlands. Some Cajuns who live along bayous and wetlands rely upon those wetlands for fishing, hunting, and trapping. Similarly, wetlands are important to several indigenous tribes in the Gulf region. The Seminole tribe of Florida, for example, depends on healthy natural resources of the South Everglades and Big Cypress regions for fishing, hunting, and economic support (conducting tours).

Sources: Gulf Restoration Network, *Wetland Importance*, http://www.healthygulf.org/our-work/wetlands/wetland-importance;National Oceanic and Atmospheric Administration, National Marine Fisheries Service, *Wetlands, Fisheries, & Economics in the Gulf of Mexico*, http://www.nmfs.noaa.gov/habitat/habitatconservation/publications/habitatco nections/num4.htm; National Ocean Economics Program. *Coastal and Ocean Economic Summaries of the Coastal States, 2009.* http://www. oceaneconomics.org/ NationalReport. For information on disaster assistance for commercial fisheries in light of the spill, see CRS Report RL34209, *Commercial Fishery Disaster Assistance*, by Harold F. Upton.

Coastal Wetlands and Assets in the Gulf of Mexico

The U.S. portion of the Gulf of Mexico has 1,631 miles of coastline.[7] There are an estimated 15.6 million acres of wetlands in both saltwater and freshwater habitats in the coastal watersheds of the five U.S states around the Gulf. (See **Figure 1**, which identifies wetlands in Louisiana, Mississippi, Alabama, and Florida.) Saltwater habitat wetlands, which occur in tidal areas where saltwater and freshwater mix, total 3.5 million acres and are of three types: estuarine intertidal emergent wetlands (salt and brackish water marshes), estuarine shrub wetlands (mangrove swamps and other salt-tolerant woody species), and estuarine and marine intertidal non-vegetated wetlands (such as sand bars, shoals, and sand spits). Freshwater habitat wetlands total 12.1 million acres and include forested and shrub wetlands, and emergent wetlands.[8] These types of wetlands generally contain shallow water and are dominated by herbaceous plants. Areas directly adjacent to the Gulf shoreline that are potentially exposed to oil from the Deepwater Horizon spill contain both saline and brackish tidal wetlands and tidal freshwater wetlands.

Environmental Sensitivity Index (ESI) maps prepared by the National Oceanic and Atmospheric Administration (NOAA) provide a concise summary of coastal resources that are at risk if an oil spill occurs nearby. In these maps, areas are scored from 1 to 10, and under this system most Gulf coast wetlands are ranked at 10, the most sensitive, or at risk.[9]

Gulf Coast Protected Lands

The Gulf coast is the site of a number of national wildlife refuges, national parks, and national seashores that are home to numerous birds and other wildlife. Four national wildlife refuges were in the path of oil as it spread to shore: Breton and Delta, both in Louisiana, and Grand Bay and Bon Secour, in Alabama); Gulf Islands National Seashore (Florida and Mississippi) also was affected.[10] Although oil has stopped flowing and the well has been capped and sealed, other refuges and National Park units could be at risk depending on fate of oil that remains in Gulf waters, and on impacts from future hurricanes and other storms. In addition to federally protected lands, many state parks and public beaches are located along the Gulf coast, as are several Department of Defense installations throughout the region that provide wildlife habitat.

Historic Loss of Gulf Coast Wetlands

Coastal wetlands are subjected to stressors from land-based activities, including dredging and filling, and from seaward events, such as coastal storms, tidal surges that cause erosion, saltwater intrusion, and inundation. In recent years, Gulf coast wetlands have experienced acreage losses from these multiple stressors. Between 1998 and 2004, coastal wetlands in the five Gulf states declined by an estimated 371,000 acres (2.3% total); although there were small wetlands gains in portions of the region, losses far outweighed the gains. Saltwater wetlands declined overall by nearly 45,000 acres (about 1.2%) during that period, while freshwater wetlands declined by 329,000 acres (2.7%). The changes in wetland acreage were much greater for Gulf coastal wetlands than for coastal wetlands in other regions, such as the Atlantic coast, which experienced a 0.1% overall decline, or the Great Lakes, which had a 0.3% increase between 1998 and 2004. The losses of wetland acreage in the Gulf coast are attributed to a number of factors. Most saltwater wetland losses in the Gulf resulted from inundation or flooding from storms or sea level rise and subsidence. Losses of freshwater wetlands have resulted from development in rural and urban areas, timber harvesting activities, drainage, and filling for extensive residential and commercial development activities.[11]

Wetland Loss in Louisiana

In Louisiana specifically, wetland change and loss have been especially severe for decades, even before the Deepwater Horizon oil spill. Wetlands are vital to many parts of the state's economy, yet they have been converting from land to water more rapidly than elsewhere. Louisiana's wetlands today represent about 40% of the wetlands of the lower 48 United States, but about 80% of historic losses. Louisiana lost about 1,900 square miles of coastal lands, primarily marshes, from 1932 to 2000, plus 200 square miles or more from hurricanes in 2005.[12] While the rate of wetlands loss has decreased, it is estimated that Louisiana loses about 25 square miles (16,000 acres) of wetlands annually, and the state experiences about 90% of the total coastal marsh loss in the lower 48 United States each year.[13]

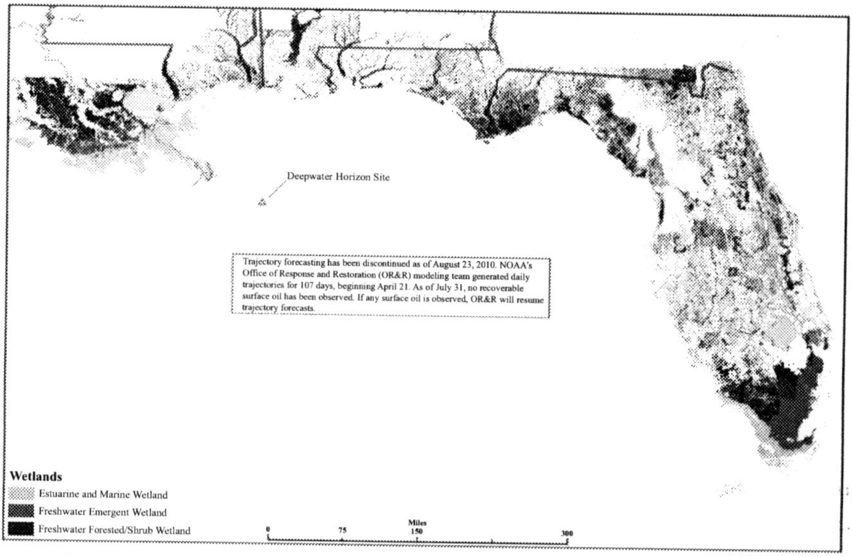

Source: Prepared by CRS based on U.S. Fish and Wildlife Service National Wetlands Inventory; ESRI Data and Maps 9.3.1.

Figure 1. Gulf of Mexico Coastal Wetlands.

Changes to the state's coastal area result from a combination of natural environmental processes (erosion, saltwater intrusion into fresh systems, sea level rise) and human-related activities, according to the U.S. Geological Survey. Wetland loss has occurred naturally for centuries, but until recently, land losses have been counterbalanced by various natural wetland-building processes.

The seasonal flooding that previously provided sediments critical to the healthy growth of wetlands has been virtually eliminated by construction of massive levees that channel the river for nearly 2,000 kilometers; sediment carried by the river is now discharged far from the coast, thereby depriving wetlands of vital sediment. In addition, throughout the wetlands, an extensive system of dredged canals and flood-control structures, constructed to facilitate hydrocarbon exploration and production as well as commercial and recreational boat traffic, has enabled salt water from the Gulf of Mexico to intrude brackish and freshwater wetlands. Moreover, forced drainage of the wetlands to accommodate development and agriculture also contribute to wetlands deterioration and loss.[14]

Also important to the entire coastal ecosystem of the Gulf are barrier islands that are a first line of defense acting as a buffer to reduce the effects of ocean waves and currents on estuaries and wetlands. Barrier islands result from the accumulation of sediments deposited at the mouths of rivers such as the Mississippi in Louisiana and the Apalachicola in Florida. Over thousands of years, sediments accumulate as unconsolidated sands and muds and gradually become compact. At the same time, marine coastal processes of erosion and degradation work at the margins of the land, eventually resulting in formation of low-relief islands that are separated from the mainland by shallow bays and lagoons. Plants often get established on barrier islands, especially on the leeward side, and help to provide stability.

But in these low-lying areas, even a moderate wind can raise water levels at the shoreline. Today, barrier islands, like wetlands, are eroding as a result of storms, sea level rise, and human development. As they disintegrate, they expose the system of sheltered onshore wetlands to additional wave action, saltwater intrusion, storm surge, sediment transport, and contamination by oil that is now circulating in the waters of the Gulf. According to the USGS, the potential exists for water to move across the full width of the islands in locations that are both low and narrow, possibly transporting oil inland into back bays and marshes.[15]

Coastal Habitat as Home

The coastal estuaries, marshes, and beaches of the Gulf also serve as habitat for a wealth of plants and animals.[16] They are among the most productive of all ecological communities. The plant life consists of plankton, algae, and larger plants, including grasses and some salt-tolerant trees such as bald cypress and tupelo gum trees. Water in such habitats is full of silt, and easily stirred by minor wave action. The turbid waters are full of microscopic single-celled algae, and other small photosynthetic microorganisms. These plants in turn feed zooplankton—microscopic animals that are food for fish larvae, juvenile crabs and shrimp, and on up the food chain. The high productivity of coastal wetlands leads to enormous species diversity, with species adapted to their own niches varying in salinity, turbidity, light penetration, wave action, and other factors that may or may not be well understood. Not surprisingly, the rich marshes and, to a lesser extent, beaches of the Gulf provide breeding grounds, nurseries, stopover points, and homes for many species, including many that are threatened, endangered, or economically important.

Breeding Grounds and Nurseries

While many commercial species are caught in deeper water, many commercial fish and shellfish depend on marshes and estuaries of the Gulf for some part of their life cycle: blue crab, shrimp, oyster, menhaden, black and red drum, flounder, striped mullet, and others. Adults of these species generally move into the marshes and estuaries, spawn, and leave eggs to hatch and develop. The eggs and juveniles are eaten by crabs, minnows, gobies, and other small predators. In turn, small alligators, herons, egrets, turtles, and other species feed on these animals, particularly where a low tide may concentrate available food resources. Insects of the marshes and estuaries feed—and many feed on—a host of young birds. Smaller birds nest in grasses, on floating vegetation, or on beaches. Larger species nest in trees, often in huge colonies. Greatest biodiversity is typically found in marshes, whether freshwater, brackish, or saline. Somewhat lower diversity occurs in open bays of fresh or saltwater, and in mudflats or marginal beaches.

The key feature of coastal ecology is that damage to the coast, and particularly to biologically rich marshes, will have effects on many species that spend the bulk of their life cycles many miles distant on land, and in oceans, lakes, and rivers. Among the most transitory of species are the migrants that use the coast as a stopover for resting and feeding during their spring or fall migrations.

Stopover Areas for Migrants

Shorebirds (plovers, gulls, terns) and wading birds (egrets, herons, ibises, bitterns), pelicans, and various species of hawks and other birds are present in the spring and fall in coast marshes. In the winter, waterfowl such as Canada, snow, and blue geese are abundant, along with dozens of species of ducks. Birds using the coast during migration may summer and breed in any part of the continental United States and Canada. Some will return to the Gulf for overwintering habitat. Others will stop in the Gulf coast area for the fall migration. These migrating birds will encounter the spill and its remaining effects just before they begin their journeys across the Gulf to winter homes in the Caribbean or Latin America.[17] In both directions, a meal immediately before or after their flight across the Gulf may be critical to reaching their destinations. Moreover, the specialized feeding behavior and anatomy of many of these species may limit their feeding sites to a narrow band of wetlands and shallow water, where oil impacts may interfere with access to foods. Consequently, depending on their migration patterns, bird populations as far away as Alaska and northern Canada or central and South America or the Caribbean may be affected by the Gulf spill.

The diversity makes Gulf coast refuges hot spots for bird watchers at all seasons, and especially in the spring migration. Of the Delta and Breton National Wildlife Refuges, which lie at the mouth of the Mississippi River, one author writes:

> Some of the most impressive sights are in the spring migration, when great numbers of small birds come in from across the Gulf, sometimes to their first landfall in hundreds of miles, weary and highly visible—rose-breasted and blue grosbeaks, indigo and painted buntings, scarlet and summer tanagers. There are as many as twenty-five warbler species, along with tremendous flights of tree swallows—sometimes sixty thousand, skimming over the inland ponds or just stopping to rest a little while before moving on north.[18]

To address the problems faced by the fall migrants, U.S. Fish and Wildlife Service (FWS, Department of the Interior) "biologists are working to restore and bolster wetland habitats and food sources in nearby, uncontaminated areas in Louisiana, the Chenier Plain of east Texas, and the lower portion of the Mississippi Alluvial Valley."[19] The effort is being carried out in cooperation with the Natural Resources Conservation Service (NRCS, Department of Agriculture), which "will be working in cooperation with private landowners and other partners to establish habitat and food sources as well as improve the overall habitat management on participating lands."[20] For the species that must feed in the immediate area of the coast, these efforts may not be fruitful, but for those birds able to feed farther inland, the additional food may be able to replace the reduced food sources at the contaminated coastline.

Threatened and Endangered Species

The Gulf coast is home to a wealth of species listed under the Endangered Species Act (ESA).[21] **Table 1** shows the total number of listed animal species in each of the five Gulf coast states, as well as the number of animal species living in that state's coastal or marine environment that could be affected by oil contamination. As the table shows, the bulk of the listed species in each state are not likely to be affected by the spill because their habitats do not occur along the coast. In a few cases, the threat to a species is conjectural, because its habitat may be fairly near the coast, but not likely to become oiled unless storms blow contamination inland. (Note that where species are at risk primarily in the event of major storms, the entry in the table adds them in by showing a range of potentially affected species.)

Listed species that are likely to have direct mortality or to suffer loss of food, nesting habitat, and the like include least terns, piping plovers, five species of sea turtles, the American crocodile, three species of whales, manatees, and three species of sturgeon. Sea turtle nests along the Gulf coast have been the focus of a special effort. As of October 12, 2010, teams of professionals and volunteers have collected 278 nests for transport to Florida's east coast where the eggs are held until the 14,676 hatching and young turtles are released, away from the spill.[22] Eventually, surviving females from these transported nests will return either to their original nest sites or to Florida's coast to lay their eggs.

In addition to the species currently listed, two species that were once listed but have recovered have also experienced effects of the spill: the American alligator, as well as the brown pelican, which has become iconic in the current case because it is designated as the Louisiana state bird. The American alligator is found fairly commonly farther inland now, but it may occur in brackish water; in the absence to date of serious Gulf coast hurricanes that can push oil inland, risk to this species has been slight.

Pelicans, on the other hand, are found almost exclusively in marine and coastal environments. Because pelicans nest in great rookeries, returning birds can contaminate their own nests as well as those of other birds. Eggs of any species that have been fouled are nearly certain to die, even with minor contamination. Chicks may be abandoned when parent birds die or become too debilitated to feed their young. Fish fed to the young may be contaminated and have uncertain effects on the birds' development.

OIL SPILLS: IMPACTS ON WETLAND HABITATS AND ANIMALS

Habitats such as salt marshes and mangrove forests and the biota that reside in them are subject to destruction or alteration by oiling events.[23] The degrees of impacts of oil on wetland vegetation are variable and complex and can be both acute and chronic, ranging from short-term disruption of plant functioning to mortality. The primary acute damage to the marshes is that plants, which hold the soil in place and stabilize shoreline, will suffocate and die, especially if multiple coatings of oil occur. Once vegetation dies, the soil collapses. Then the soil becomes flooded, and plants cannot re-grow. If plants cannot re-establish, soil erosion is accelerated, giving rise to even more flooding and further wetland loss.

Table 1. Endangered and Threatened Animal Species in the Gulf of Mexico (Coastal species in habitats that may become oiled)

State	Total Endangered Animal Species	Endangered Animals: Possibly Oiled Habitat	Total Threatened Animal Species	Threatened Animals: Possibly Oiled Habitat
Texas	49	8	9	3
Louisiana	10	8	8	4-5
Mississippi	20	10	11	4
Alabama	58	10-11	21	4
Florida	35	23-24	19	9-10

Source: Compiled by CRS from FWS Endangered Species Program Data, Species Profiles. Available at http://www.fws.gov/endangered/.

Notes: All listed coastal and marine species in a state are considered as being in potentially oiled habitat, except certain species occurring only on the east coast of Florida. A range of figures indicates that the exact location of some coastal species is not known, or that effects on the listed species might depend on the occurrence of severe storms washing contaminated oil well into coastal marshlands. Columns should not be summed, since many coastal species occur in multiple states. An FWS website list shows fewer animals, but excludes species under the jurisdiction of NMFS; see http://www.fws.gov/home/dhoilspill/pdfs/FedListedBirdsGulf.pdf.

If oil penetrates into the sediments, roots are continuously exposed to oil, with chronic toxicity making production of new shoots problematic. Consequently, plant recovery is diminished, and eventually land loss occurs. In addition to direct impacts on plants, oil that reaches wetlands also affects animals that utilize wetlands during their life cycle, especially benthic organisms that reside in the sediments and are a foundation of the food chain.[24]

In the Gulf, coastal saltwater wetlands and tidal freshwater wetlands are sensitive to oil, both immediately and in the long term. The severity of immediate and longer-term impacts and damages depend on a number of factors, including—

- the type and quantity of oil spilled (low-sulfur light crude such as that associated with the Macondo well is considered less toxic than denser heavy crude or refined fuel oils involved in other spills. Still, the quantity of oil from the Deepwater Horizon spill is the largest that has ever occurred in U.S. waters);
- the condition of the oil on and below the surface, including the length of time it is in the water before it hits land (oil weathers over time, through the natural actions of evaporation, photodegradation, and microbial degradation, which all reduce acute toxicity);
- season and prevailing weather (both the natural degradation processes and ecosystem recovery occur more rapidly in warmer climates than colder regions, but a spill that occurs during the growing season of plants will have more severe impact than one in the fall or winter when plants are dormant);
- type of shoreline and composition of vegetation (wetlands and similar fragile environments are more at risk and more difficult to clean than harder surfaces such as sandy beaches);
- type of waves and tidal energy in the area of the spill (wave energy is necessary to help move oil away from land and to enhance the effects of chemical dispersants); and
- presence of dispersants (chemical dispersants are used to enhance breakdown of the oil, but the long-term ecosystem effects of chemically dispersed oil and the dispersants themselves are uncertain).[25]

Estimating Mortality

The effects on wetland habitats lead directly to effects on the animals and plants in those habitats. The federal government has the lead in managing some categories of species, while states take the lead in the rest. Comprehensive data are more readily available for the former, which are called federal "trust species." The trust species consist of birds, marine mammals, and threatened and endangered species. The principal federal agencies involved in managing these species are the FWS and the National Marine Fisheries Service (NMFS, Department of Commerce).

Birds

Oil contamination of birds can have acute effects:

> Birds are primarily affected through thermoregulatory challenges caused by oiling of plumage and through ingestion. Oil reduces the ability of bird feathers to provide insulation, which increases their risk of hypothermia. Rate of heat loss is much higher in the water than in air, so oiled plumage is particularly problematic for birds that must find food in the water, such as seabirds, cormorants, and grebes. Oil is most commonly ingested by birds while preening their contaminated feathers or while feeding on contaminated prey.[26]

The President of the American Bird Conservancy stated that "rescue groups are prepared to do everything humanly possible to capture and save as many oiled birds as they can find, but there are problems well beyond our abilities to mitigate or even count. In addition to the potential catastrophic losses to shorebirds that we know to be at risk on their breeding grounds and in the wetlands around the Gulf, the oil spill poses a serious threat to seabirds."[27]

The coating of feathers also results in loss of buoyancy, and, if severe, can cause the bird to drown and then sink. It is this phenomenon that makes an accurate estimate of bird deaths extremely difficult. Based on past spills, many argue that mortality analyses based on counts of carcasses and on estimates of the ultimate survival of oiled birds that have been cleaned and released produce a marked undercount of actual deaths, particularly for species that forage farther offshore. According to the Exxon Valdez Oil Spill Trustee Council, after the 1989 Alaska spill, the number of recovered carcasses of birds was only a fraction of the estimated deaths:

The carcasses of more than 35,000 birds and 1,000 sea otters were found after the spill, but since most carcasses sink, this is considered to be a small fraction of the actual death toll. The best estimates are: 250,000 seabirds, 2,800 sea otters, 300 harbor seals, 250 bald eagles, up to 22 killer whales, and billions of salmon and herring eggs.[28]

The number of live but contaminated birds found in the Gulf as of October 12, 2010, was 2,080, along with 6,104 carcasses, for a total of 8,184.[29] Of the birds collected alive, 1,243 were released after examination, cleaning, and recovery. Carcasses are stored, and will be used as evidence in litigation concerning damage to natural resources. If no more carcasses were found, and if the same ratio were to apply to the BP spill as in the *Exxon Valdez* spill, then about 43,600 bird deaths would be expected.[30] However, not only does this calculation assume (improbably) that no more carcasses will be found, but also that the diligence and thoroughness in collecting carcasses in the Gulf is similar to the diligence and thoroughness in Prince William Sound.

Marine Mammals

Marine mammals inhabiting the Gulf include manatees, several species of whales, dolphins, and seals. These species do not have significant hair that can become oiled, but rely on a layer of fatty tissue for warmth. However, contact with oil can cause skin irritations, perhaps leading to infections. A more significant threat is inhalation of fumes when they surface to breathe. Moreover, their prey may be contaminated with hydrocarbons, or the prey populations may be reduced or absent. Some of the Gulf's whales feed by sifting large amounts of water through plates called baleen. These plates may become clogged with oil.[31] As of October, 12, 2010, there were 9 marine mammals collected alive, and 97 dead; of the carcasses collected, 91 did not have visible oil, and a necropsy will be needed to determine the cause of death.[32] While marine mammals may be physically capable of avoiding oil slicks, according to the National Oceanic and Atmospheric Agency, "[r]esearch on dolphins in human care has shown that the animals avoid oil on the surface of the water, however observations of wild dolphins have documented the animals swimming in, feeding in and socializing in oiled water during previous oil spills in the Gulf of Mexico."[33] Where a spill covers a very large area and volume of water as it does in the Gulf, the animal may have to avoid much of its previous range.

Endangered Species

Threatened and endangered species living in the wetlands and coastal areas face varying degrees of risk. Moreover, the threats to a given individual may be different from those to a species: a breeding bird that is unable to find sufficient food to feed its young may abandon its attempt to breed for the year, and the young will die without the care of the parent bird. Or the eggs may become oiled, fail to hatch, and be abandoned. But the adult can survive itself. In either case, if abandonment is widespread, the damage to the species could be severe even if all adult birds were to survive.

WEATHER AND STORMS

The effectiveness of offshore measures in capturing or breaking up oil in open waters of the Gulf in part determines whether oil reaches coastal areas. One factor that can influence the success of such measures is storms and weather events, since high winds and strong tides can push oil further inland from shoreline areas.

The official hurricane season in the Atlantic Ocean begins June 1 and ends November 30. On August 5, 2010, the National Hurricane Center (NHC) issued its updated outlook for 2010 hurricanes in the Atlantic and Gulf of Mexico, predicting an "active" season with "8 to 12 Hurricanes (top winds of 74 mph or higher), of which ... 4 to 6 could be Major Hurricanes (Category 3, 4 or 5; winds of at least 111 mph)."[34] If hurricanes occur while large amounts of oil are present in Gulf waters and at shorelines, several concerns emerge. First, barrier islands that normally protect coastal wetlands from oil may be overtopped or destroyed by winds and water. Second, heavy winds and storm surges could push oil further inland into areas that would not normally become oiled. Third, even though the oil stopped flowing from the Deepwater Horizon site in July, hurricane-strength winds could stir up oil that has penetrated sediments, thus releasing oil and causing renewed impacts on plants and wildlife. Fourth, tropical storms and hurricanes can disrupt or delay ongoing oil spill response and cleanup activities: this has been the case resulting from several storms in June and August that affected Gulf waters and shores.

According to the NHC, the presence of oil is not expected to affect either the intensity or the track of a fully developed tropical storm or hurricane appreciably, and an oil slick in and of itself would have little effect on a hurricane, a storm surge, or nearshore wave heights. In addition, high winds and seas generally mix and weather the oil, which actually can help accelerate biodegradation processes and disperse the oil further. Movement of oil would depend greatly on the track, wind speed, and size of a hurricane, and other factors in the evolution of the storm.[35]

Mitigation and Cleanup of Wetlands

Much of the response effort in the Gulf has focused on capturing, or otherwise addressing, oil in open waters or setting barriers to deflect it from reaching land. Protective booms placed offshore were intended to prevent oil from reaching shore (most were removed after oil stopped flowing from the well). There is evidence that these and other measures have been at least partially successful in reducing the volume of oil and preventing oil in very large amounts from contacting marshes and beaches, as had been anticipated and feared initially. Another defensive measure, a project to build sand berms to provide additional physical protection for Louisiana's barrier islands, is being constructed, despite controversies about the effectiveness of the approach.[36]

But, inevitably, some oil escapes capture and is pushed by wind and tides toward land. Wetland plants can be affected both by oil that floats over the surface of the marsh and by oil that has been incorporated into sediment. There are several possible approaches to cleaning up oil that reaches coastal wetlands and competing theories of different approaches in different places. Moreover, experts acknowledge tradeoffs for each approach. Each has limitations—some of which are serious—and none is effective in all cases. As Admiral Allen said, "The problem is, the hardest place to pick up oil is in a marsh or a wetland."[37] There is no silver bullet and no single best solution, and often the response can be more destructive than the spill itself. Some studies even suggest that a certain level of oil in the soil of wetlands may stimulate plant growth, although the mechanisms by which this occurs are not clearly understood.[38]

Mechanical Recovery

Mechanical removal techniques are often used first, once oil reaches land. It is possible in some cases to physically remove oil by hand, with shovels or rakes, vacuums, or other equipment, or to skim and mop it with absorbent materials. However, these methods are difficult, and their effectiveness is highly variable. This technique works best on sandy beaches or rocky shores but is difficult in marshy areas, because damage from heavy equipment and human trampling of sensitive plants and soils can equal or exceed damage from the oil. Recovery of wetland vegetation from the mechanical cleanup can take considerable time. Further, oil sitting on the surface of shallow waters can get pushed into soft ground where, in the absence of oxygen, it contaminates sediments or biodegrades very slowly.

Flushing

In some cases low-pressure flushing can be used to push oil into areas where it can be vacuumed or absorbed. Seawater is pumped through the marsh to dilute the oil that sticks to grasses, thus enhancing tidal movement to promote natural recovery. High-pressure spraying can be effective. It was used to clean the rocky shore of Prince William Sound in Alaska, although with some negative effects on shoreline invertebrates, but the Gulf coastal wetland environment is very different from that in Alaska, and high-pressure spraying likely would erode soils and destroy wetland vegetation. Even low-pressure spraying only works when oil is floating on the surface.

Applying Chemical Dispersants

A related technique is to spray dispersants—a mixture of surfactants, solvents, and other chemicals—onto the oil. Like dish soap, dispersants break the oil into smaller droplets that fall from the surface into the water column, where microbes degrade them or currents move them away. This technique is primarily used in open waters: chemical dispersants have been used in the Gulf spill to break up oil before it reaches shorelines.[39] However, there is some evidence that when dispersants are added to water during flushing of marsh vegetation, oil removal is slightly enhanced.[40]

But there are many questions about environmental effects of chemically dispersed oil and the dispersants themselves in wetlands. Many scientists contend that adding dispersants enhances removal and that on balance the use of dispersants is less harmful than letting oil get into marshes. However, other scientists are concerned about harmful effects of dispersants on aquatic plant and animal life in both open waters and coastal areas. If dispersants are used, there are environmental tradeoffs: (1) greater concentrations of chemically dispersed oil in the water column, (2) a potential reduction in persistent stranded oil, and (3) unknowns in terms of long-term toxicity.[41] Long-term effects of applying dispersants to oil in coastal wetlands are not well known, because research has focused more on acute toxicity than chronic effects. If dispersant chemicals penetrate sediments, the toxic effects might be long-lasting.[42]

Burning

Burning is considered one of the most effective techniques for responding to spilled oil. It was used successfully to clean up several oil spills following Hurricane Katrina, and, when wind conditions permit, it has been used *in situ* in the Gulf to prevent oil from reaching shore.[43] However, there are limits to how and when burning can be used. Burning coastal plants may kill them and result in further coastal erosion. It can be difficult to control the burn and confine it to small areas, thus threatening other plant material not affected by oil. Burning removes oil's more toxic components, such as volatile aromatic compounds, but it does not destroy stickier, heavier (high molecular weight) compounds that can sink into wetland soil. There is evidence that the concentration of oil in sediments is greater in areas that have been burned than in unburned areas.[44] Marshes with low water levels cannot be burned without killing the plant roots. But research has shown that the presence of just a few inches of water above the roots when a burn occurs can be sufficient to provide sufficient protection, thus allowing subsequent regeneration of the plants.

Also, oil that is weathered, or degraded (that is, petroleum that has been on the surface and exposed to air for several days), does not burn well, because flammable hydrocarbons have volatilized. That has likely been the case with oil from the Deepwater Horizon spill, as it typically took more than a week for tides and winds to move oil from the rig site to shore.

Cutting Back Vegetation

In small heavily oiled areas, vegetation can be cut back to leave plants intact and prevent oiling of birds that come in contact with the vegetation. This technique is moderately effective at cleaning up oil, and it has been successfully used in the past for smaller spills and plant regeneration has followed (including spills affecting Gulf coast marshes in the past). But it does not work well in large areas. One of the risks is physical damage from human contact and trampling. Some experts believe that areas where grasses are cut take much longer to recover than oiled areas that are not cut.

Bioremediation

Biodegradation is the process in which oil molecules are broken down by bacteria, which occurs naturally. However, to make oil biodegrade faster, bioremediation agents or techniques are sometimes applied. Bioremediation involves either introducing nutrients in the form of fertilizer into the marsh (i.e., biostimulation) or adding bacterial microbes designed to be especially effective at degrading oil (i.e., bioaugmentation). EPA has made information available about dispersants and bioremediation agents that are authorized for use on oil discharges in the United States.[45]

This technique can be effective in some places, but where it has been used on spills in wetland areas, results have been mixed. Experts believe that it probably has limited potential in coastal marshes such as those in Louisiana, because oxygen levels in wetland soils often are so low that microbial activity is limited, whether nourishment is added or not. That is, oxygen is the factor limiting degradation of oil, not nutrients, and there are no proven methods of adding oxygen to muddy, water-saturated marsh soils. It is likely to be more effective in wetlands at a higher elevation than coastal Louisiana, or areas that are infrequently flooded, because more oxygen is present. Bioremediation is not usually considered as a first response but is more often a secondary treatment after removal of bulk oil.

Nutrient addition may enhance the growth and productivity of dominant plant species within an oiled area, but it also may reduce species diversity or composition among remaining wetland plants. It may even be toxic to some organisms. Bioremediation in combination with detergents was used during cleanup of the *Exxon Valdez* spill. In that instance, treated areas were significantly cleaner after the first year, but scientists later determined that the addition of large amounts of fertilizer or nutrients disturbed the ecological balance of the overall system, thus altering the course of long-term recovery.[46]

Doing Nothing

It may seem counterintuitive, but many scientists believe that doing nothing is sometimes the best option, contending that aggressive cleanup or remediation can have serious harmful effects and may delay the eventual recovery that might naturally occur. According to the National Research Council, there is no consensus on whether it is better to immediately clean up an oiled area that may then take many years to re-establish, or leave it alone to weather naturally and risk uncertain effects over long periods. In many instances, cleanup techniques delay recovery time, due to physical disruption of roots, flushing of soils, lowering the soil surface below levels where aquatic vegetation can re-establish, and activities that mix oil deeper into wetland and marsh soils.[47]

On the other hand, if oil is allowed to persist around plants, it kills the existing vegetation and can prevent new shoots from growing when they contact floating oil, which could lead to more wetland erosion. By deciding to do nothing, there are essentially no direct environmental side effects, and it may still be possible to explore alternatives later. But several considerations do apply: when the oil is especially heavy (although that is not the case in the Gulf spill), or if tides are insufficient to move the oil, unassisted or natural recovery may not be completely effective. Also, doing nothing may be more appropriate for a one-time event (such as the *Exxon Valdez* incident) than for a spill in which oil continued to flow for weeks or months, as with the Deepwater Horizon spill. Further, because effectiveness may not be known for months or years, it can be difficult to persuade the public that doing nothing is best.

OIL SPILL RESPONSE: WHO DECIDES WHAT TO DO?[48]

Several options, including technologies, exist for responding to oil that reaches coastal wetlands, but all have apparent advantages and disadvantages, as previously described. Once the oil moved from the open Gulf waters to shoreline areas, decisions about how to actually respond were made within the overall complex organizational structure that exists in the United States for responding to oil spills.

The federal government's oil spill response framework is found in the National Contingency Plan, which contains the government's procedures for responding to oil spills and hazardous substance releases. The National Oil and

Hazardous Substances Pollution Contingency Plan (NCP) was established administratively in 1968, after U.S. policymakers observed the response to a 37-million-gallon oil tanker spill (*Torrey Canyon*) off the coast of England and saw the need for a coordinated approach to cope with potential spills in U.S. waters.[49] Subsequent laws have broadened the NCP, including the Clean Water Act in 1972 (33 U.S.C. 1251 et seq.); the Comprehensive Environmental Response, Compensation, and Liability Act (CERCLA or Superfund, 42 U.S.C. 9601 et seq.) in 1980; and the Oil Pollution Act (OPA, 33 U.S.C. 2701 note) in 1990.

The NCP establishes the National Response System (NRS), which is intended to be a multitiered and coordinated national response strategy. Key components of the NRS include:

- a National Response Team (NRT), composed of representatives from the 15 federal departments and agencies assigned roles in responding to oil spills.[50] The U.S. Coast Guard chairs the NRT when a response is being mounted to a spill in a coastal region.
- Regional Response Teams (RRTs), composed of regional representatives of each NRT member agency, state governments, and local governments. The Coast Guard leads the relevant RRT during responses to oil spills in coastal waters.
- Area Committees (ACs), composed of qualified personnel from federal, state, and local agencies. The primary function of each AC is to prepare an Area Contingency Plan (ACP) for its designated area.
- an On-Scene Coordinator (OSC), who directs the response efforts and coordinates all other efforts at the scene. In general, in coastal areas Coast Guard Captains of the Port serve as OSCs for their particular area, but in the current case, the federal OSC is Rear Adm. Paul Zukunft, who is the Coast Guard's assistant commandant for marine safety, security, and stewardship.

The NCP provisions specific to oil spill response are found in 40 CFR Part 300, Subpart D. As the primary response authority in coastal waters, the Coast Guard OSC has the ultimate authority to ensure that an oil spill is effectively removed and actions are taken to prevent further discharge from the source. The OSC is broadly empowered to direct and coordinate all response and recovery activities of federal, state, local, and private entities (including the responsible party), drawing on resources available through the appropriate ACPs and RRTs.

To manage response operations for the Deepwater Horizon spill, the federal government established a Unified Command structure, led by a National Incident Commander (Coast Guard Admiral Thad Allen in the current case), to link the organizations responding to the incident and provide a forum for those organizations to make consensus decisions.

Although the OSC must consult with designated trustees of natural resources regarding damages[51] and with the governor(s) of the state(s) affected by the spill, the OSC has the authority and responsibility to determine when removal (i.e., cleanup) is complete.

As oil from the Deepwater Horizon spill reached the shores of several Gulf coast states, the entities charged with responding to the spill began to focus on cleanup of coastal marshes and wetlands, particularly on considerations related to the several possible cleanup techniques discussed in this report. In early June, representatives of several federal agencies that are part of the NRT, plus emergency response and local technical experts, held a workshop to develop a response plan for marshes and other nearshore and shoreline areas affected by the oil spill. The resulting plan identified current cleanup methods and strategies along with their advantages and disadvantages, depending on the level of oiling of marshes, and included a table showing potential activities that could be considered by the Incident Commander (Admiral Allen).

As to which methods are actually to be utilized, the response plan stated, "The selection of any cleanup strategy will be determined on a case-by-case basis in consultation with appropriate natural resource experts and trustees. Clean up operations will be conducted by the responsible party and overseen by the Federal, State and Tribal Agencies with authority."[52]

However, while the oil was still flowing, very little cleanup actually was occurring, at least near or on the shore, since recontamination could occur readily. Rather, under the National Incident Command structure, Shoreline Cleanup Assessment Teams (SCATs) undertook comprehensive surveys and collected information on oil once it reached shore.[53] Information that the assessment teams reported daily—and continue to report even after oil from the well stopped flowing—is processed by the Incident Command Center as part of broader planning and priority-setting to guide operations. Shorelines have been resurveyed multiple times, as oiling conditions changed, and some cleanup recommendations likely changed as a result. Until oil stopped flowing in mid-July, shoreline cleanup was limited to triage—removing bulk amounts of oil from beaches and, in some cases, heavily oiled marshes—but even these areas were not being thoroughly cleaned. Methods being used in the marshes were primarily flushing and sorbents (i.e., absorbent booms). Given the high environmental sensitivity of marshes and wetlands and the potential for re-oiling until the leaking undersea well was capped, the initial choice was to do as little as possible that could be more harmful to the ecosystem.[54]

With the flow of oil from the well apparently stopped, and the emergency phase of the spill response over, more comprehensive cleanup continues. The biology of an area or type of use will determine the desired cleanup endpoint. That could mean, for example, zero tolerance for oil on a high-amenity recreational beach and nearly the same for a marsh. A number of questions can be anticipated to arise during this process.

- What is meant by "case-by-case" determination of cleanup strategy, in the NRT response plan? That is, what factors (physical, biological, legal), would result in different methods being used at some sites? Would this decision include scientific controls to decide what methods are most effective?
- What kind of training is being provided to persons who are actually involved in cleanup? While many persons are reportedly being trained and used for cleanup of less complex areas, such as beaches, different preparation may be required for those who will perform cleanup work in sensitive marshes and wetlands.
- The NRT response plan necessarily is focused on removal of oil, which likely will continue for some time. But even in the early stages, there have been opportunities for scientific evaluation arising from these events. How are the needs to improve scientific understanding of the impacts of the oil spill and response activities being considered during cleanup?

Who Decides about Wildlife?

The process of cleaning an individual animal may take a week or more: collecting and transporting the animal; rest and/or feeding and hydration to allow the animal to withstand the stress of the cleaning or rehabilitation process; cleaning and treatment; drying; further feeding and blood tests, and rest; transport to a release site; and release, often at a very distant location.[55] As with many wildlife questions, data are better for birds than for many other species.

Cleanup of wildlife is controversial, in terms of both the benefits to individual animals and the question of the cleanup's relevance to the status of the species. The issue is, if only a small fraction can be saved, and if the food sources, breeding sites, and resting places are damaged or destroyed, how useful is it to clean a small fraction and release the animals to an environment where they may become recontaminated? Even if one focuses on the individual birds rather than the species, some scientists argue that the life expectancy of a bird that has been cleaned and released remains low, and its chances of breeding are poor, although some groups, such as gulls, fare better than other groups.[56]

There may be no definitive answer to such questions, other than the human need to do something that might help, plus the expectation that at least some fraction of the cleaned animals will survive at least somewhat longer. To those who hold this view, the rescue of oiled animals is not amenable to a simple cost-benefit analysis. It may be no coincidence that much of the rescue of oiled animals is done by dedicated corps of volunteers.

Cleanup and Recovery Issues: The Long Term

Once the oil is removed or cleaned up to the extent possible, recovery of the ecosystem can begin. Recovery can occur within a growing season or require years to decades, based on factors such as those listed previously that influence the severity of impacts (e.g., the type and quantity of oil, and season and prevailing weather). Documented recovery of oiled marshes is more rapid in the warmer environment of the Gulf than in colder climates, but it still ranges from weeks to decades.[57] Lessons from the *Exxon Valdez* spill may be useful in the BP oil spill for both the biological response and for management of scientific information.

Lessons from the *Exxon Valdez*

The *Exxon Valdez* spill in 1989 had well-known short-term effects, but its long-term effects are less well known. While the markedly different conditions (a single event with short-term oil release; much colder climate; preponderance of rocky shores over marshes, beaches, and mudflats; lower human population; fewer long-established natural oil seeps and therefore fewer native microbes capable of digesting oil; less disturbed baseline ecosystem, etc.) mean that many lessons do not transfer well, there are two lessons particularly worth noting as Gulf onshore cleanup plans develop.

First, short-term remediation actually delayed recovery in some habitats. For example, while rocky shores cleaned with high-pressure hoses or hot water became suitable for the hauling out of seals or sea otters, other aspects of recovery were negatively affected: microbial populations that might have digested oil were destroyed.[58] At these sites, the recovery process for intertidal species (barnacles, crabs, limpets, etc.) proceeded more slowly than in untreated areas in which some oil weathered and some sank into the rocks, sand, or gravel.[59]

Second, before the spill in Prince William Sound (PWS), the paradigm for oil spills had been that a spill's effects were primarily acute, rather than long-term. But in the intervening 21 years, studies have shown that some effects can endure for decades, and that some species recover much more slowly than others. Research did not continue indefinitely after the spill, though some responses were studied for longer than others. The following list gives a sample of the conclusions of a scientific review of the published information on ecosystem responses in the years after the spill:

- While about 2% of the oil mass remained on PWS intertidal beaches after 3.5 years, the rate of decay of this oil dropped markedly in subsequent years because it became sequestered from wave action, sunlight, and oxygen—all of which tend to promote degradation of the oil, as previously described. About 14% of the oil that contaminated PWS tidal areas in 1992 was still present in 2001.
- Where mussels anchored themselves to these rocky shorelines, the subsurface oil contaminated the mussels and provided a route into the food chain.
- In the mouths of rivers, oil in the substrate killed pink salmon embryos at least as late as 1993 (the most recent year available to the reviewers).
- Sea otter populations recovered at the rate of 4% per year, short of the expected 10% per year. By 2000, sea otter populations remained at about half the estimated pre-spill population in one heavily contaminated area. Effects included an abnormally high mortality in animals born after the spill.
- Some birds, such as harlequin ducks, showed higher mortality rates than normal for more than a decade. This species feeds heavily on invertebrates in lower tidal areas. Another sea duck, Barrow's

goldeneye, also showed chronic exposure for at least a decade. This species also forages in the nearshore environment.

- Sublethal effects occurred in other species. For example, black oystercatchers (a bird that forages on rocky shores) had less breeding and smaller eggs for at least three years.
- Among social species, such as killer whales, loss of key individuals can affect large numbers of animals: when adult females in this matriarchal species died after the *Exxon Valdez* spill, the social disruption apparently led to suppressed reproduction, and likely to the later disintegration of a pod of whales.[60]

While most of the species mentioned above are not found in the Gulf, in both the Gulf and PWS, there exists the opportunity for hydrocarbons to become lodged in various substrates, to be absorbed by various species that are eaten by other species, and to inhibit the growth or reproduction of still others. It remains to be seen how different the biological response in the Gulf will be, but the *Exxon Valdez* spill suggests that long-term impacts are highly likely, even if their specific nature will be difficult to predict.

Moreover, the resulting effects of the *Exxon Valdez* spill could be compared scientifically to the effects of the BP spill in new research. The diverse habitats, wildlife, microbes, shoreline structures, etc., could provide insight on how to handle future spills in still more environments, as oil development moves into more challenging areas.

Science and Litigation

Free and open access to scientific information concerning oil spills is not a given. In the Gulf, many lawsuits have been filed already and more can be expected. In the past, litigation strategies have driven parties in lawsuits to direct government or corporate scientists to withhold data in order to prevent opponents in court from having an advantage as litigation progressed. After the *Exxon Valdez* spill, the Alaska attorney general:

> issued a series of memos to state scientists ordering them to keep their data on the spill under wraps. His motivation, [he] said, was to prevent Exxon from gaining the upper hand in the litigation that was shaping up from the moment the *Valdez* ran aground.... The gag has been particularly effective in muzzling the most contentious data—that relating to environmental damage....

> Indeed, far too little data has been made public for the response experts to come to a unanimous opinion of the best strategies for cleanup. In some cases, the response specialists have not been allowed to see the state's damage assessment results....[61]

In the wake of that spill, contractors working for Exxon were allowed to release results of studies that concluded that the spill had not had, and would not have, adverse effects on plants and animals in PWS. Alaska's scientists could not provide timely contrary information.[62] At the same time, other Exxon researchers were not allowed to present their research on the spill's effects, due to the possibility of harming Exxon's defense.[63]

The tension between disclosure of scientific research and litigation strategy has already begun in the Gulf, with independent scientists complaining that, unless they are employed by BP, or by the federal or state governments, they are not allowed access to coastal areas in a manner that would permit independent research.[64] In addition, various organizations, including the National Wildlife Federation and the American Association of University Professors, voiced concern that BP asked a number of scientists in the region to do research on the spill, and to sign confidentiality agreements lasting three or more years as a condition of doing so.[65]

Confidentiality agreements have been signed by many federal employees, contractors, subcontractors, or other persons hired in connection with the employee's work on the spill. The terms of the Department of the Interior (DOI) version of these two-page agreements include these provisions:

> All documents, information or other work developed by me in connection with this matter are privileged and confidential. I shall not reveal any of these documents, information or other work to any person, other than employees, subcontractors, or persons that I hire to assist me in connection with my work on this matter, without prior written approval of the office of the Solicitor, United States Department of the Interior....
>
> Unless prior written approval of the Office of the Solicitor, United States Department of the Interior is obtained, the undersigned shall not engage in publication or public discussion of confidential material, information, documents, or work product in any forum, scientific, technical or otherwise, prior to the completion of a natural resource damages assessment and final resolution of any and all administrative or judicial actions related to this matter.[66]

In one respect, the agreement differs from the agreements reportedly signed by BP contractors: federal agreements have (or had) no specific time limits. Release is contingent upon final resolution of relevant administrative and judicial actions. If another spill were to occur in the Gulf or elsewhere, parties to a later spill would have no clear way of knowing whether research on the effects of the BP spill would be useful in forming a response, unless a federal employee were aware of the research and attempted to clear it with the DOI Solicitor for use in responding to the later spill. At the same time, high ranking federal officials claim to be unaware of restrictions on independent researchers or on their access to study sites, unless the researcher wishes to participate in NRDA formally.[67]

When the scientific community voiced its concerns about such requirements, BP reportedly changed its policies.[68] In addition, it committed $500 million for a fund (Gulf Research Initiative, or GRI) for competitive grants for scientific research. According to a BP press release "All GRI-funded research will be independent of BP, and the results will be published in peer-reviewed scientific journals with no requirement for BP approval."[69] NOAA is also reported to have relaxed its restrictions on scientific publication.[70]

It appears that the state of scientific research, access to study sites, and rules for publication are in flux. There was clearly no default setting for science research among federal agencies or independent scientists at the start of the oil spill, although there is some indication of greater freedom to do independent research. A general policy for determining what science can be done, and what access researchers may have, particularly during the critical immediate response to crisis, may benefit responders to future crises. The Gulf oil spill might also produce broader agreement on when and how research resulting from a disaster may become generally available.

CONCLUSION

The long-term impacts from the Deepwater Horizon release on the environment will be investigated for some time, probably many years. There are unanswered questions about the amount of oil still remaining in the Gulf, how the oil will degrade in the environment, what mitigation actions will be taken, and what immediate and long-term damaging effects might result from those actions.

Oil has the potential to endure in the environment long after a spill. In coastal marshes, oil may be pushed or may seep into bottom sediments and may persist for a long period of time, having potential to be re-suspended by waves or storm events, and remaining a threat to wetlands and wildlife.

The costs of the oil spill in damages to wildlife and wetland resources—killed or injured animals, contaminated or destroyed vegetation and habitat—cannot be measured at this point. Likewise, the financial costs of response and recovery will be unknown for years. Those financial costs will include public and private response activities, fines and penalties, and third party claims. They also will include natural resource damage settlements. In the current case, natural resource damages and assessment costs—including those for determining the extent of impacts, restoring injured resources to baseline, and compensating for interim losses—are likely to be substantial in absolute terms, but still could be small compared with response costs and third party claims.[71]

Although events are still at a relatively early stage, a number of questions arise. When will recovery be complete, and how do we define success, or completion? What is an acceptable level of habitat recovery? NOAA regulations (15 C.F.R. § 990.30) state that recovery "means the return of injured natural resources and services to baseline"—in other words, a return to conditions as they would have been had the spill not occurred. But what this means in actual terms is open to considerable debate. A conventional definition of recovery is probably "return to the way things were before the spill." Unfortunately, this benchmark is vague and hard to quantify. Change is particularly characteristic of coastal areas, which are already subject to significant annual losses in the Gulf coast area. In the case of Prince William Sound, for example, it is difficult to apply this standard of recovery because there is little information about "the way things were" before that spill. Multiple variables affect local species and ecosystem services. Similarly, one species at a spill site could have been on the decline at the time of an incident, because of changing water temperatures or other factors. Even in the absence of a major disturbance like an oil spill and cleanup, ecosystems are constantly in transition, and the physical and biological conditions that once characterized any given site are likely to shift considerably over time.[72]

With so many unknowns, predicting the long-term future of the wealth of natural resources in the Gulf could be foolhardy. Even so, observations from the *Exxon Valdez* spill and other spills offer some insight. First, there will likely be significant long-term effects, and the impacts will not be evenly distributed. Some areas and species will require more time than others. Moreover, the stresses of these effects are an add-on to wetlands that are already disappearing due to subsidence, global sea level rise, and other factors. If a wetland area simply disappears as a result of some combination of these factors, no recovery can occur, regardless of remediation. The Gulf's wildlife was already under stress from these same factors. The spill seems likely to exacerbate existing trends, although in ways that are hard to predict.

Second, litigation has been a driver of post-spill action in the past, and multiple lawsuits have already been filed in response to the Deepwater Horizon event. If the *Exxon Valdez* experience is an indicator, the timely sharing of data among scientists appears likely to suffer. Some information could be temporarily or indefinitely concealed, possibly to the detriment of recovery of various resources. Since much is unknown about coastal wetlands and their inhabitants, the withholding of data could be a handicap to recovery.

Third, the spill did provide at least one benefit that might not have occurred otherwise. Federal employees from many agencies have reported for weeks of duty, sometimes as volunteers, to use their expertise in responding to the spill.[73] They, and a large number of state and local employees, as well as volunteers, now have experience in responding to a disaster of this magnitude. As a result, a large number of government employees and others have had training in the Incident Command system; this training could be useful in future disasters of any nature.[74]

When immediate response to the spill and recovery (however it is defined) are complete, attention will likely turn to restoration. At that point, another question that will arise is, what does restoration mean? In June, President Obama committed the federal government to a long-term plan of Gulf coast environmental restoration. The White House said that a federal interagency task force, led by Navy Secretary Ray Mabus, will develop a plan with the goal of making the Gulf better than it was before April 20 (or, better than the baseline condition called for in NOAA's regulations that define recovery). When that plan emerges, it will represent the most recent such proposal, adding to numerous other plans and projects that have been proposed to rebuild coastal areas that have experienced decades of adverse effects from erosion, storms, and development, especially in Louisiana's wetlands. Some of those earlier plans were endorsed by Congress,[75] and some of them are underway now, but critics observe that there has been little action to date for reasons that include costs, complexity, and ecological and institutional challenges. Of great interest will be how existing and new plans and proposals will be coordinated, how current short-term responses to the oil spill may affect long-term restoration,[76] and how barriers that have stalled previous efforts can be overcome.

During the immediate response to the spill, Congress has provided substantial oversight of efforts to contain and mitigate the oil and its impacts.

Congress also is considering policy and legislative changes to prevent such events in the future.[77] In the longer term, Congress is likely to be influential in guiding the environmental restoration plan for the Gulf coast that the President has pledged to pursue. During the immediate response to the oil spill, two key questions have been, who is responsible for cleanup, and who pays.[78] Similar questions will arise in relation to longer-term restoration, especially questions of who will pay for restoration and what the federal role will be.

End Notes

[1] Adm. Thad Allen press conference call, May 21, 2010, p. 7, http://www.deepwaterhorizonresponse.com/posted/2931/ Teleconference_Allen_May_21.562827.pdf. Adm. Allen retired from the U.S. Coast Guard in early 2010, but continued to serve as National Incident Commander until Oct. 1, 2010.

[2] See http://www.restorethegulf.gov/release/2010/05/07/breton-national-wildlife-refuge-closed-public-entry, reporting closure of Breton National Wildlife Refuge (LA), reporting shoreline contamination on May 5, 2010. For frequently updated maps of the spill, response, fishery closures, and other geospatial features, see http://gomex.erma.noaa.gov/erma.html.

[3] Unified Command for the BP Oil Spill, "The Ongoing Administration-Wide Response to the Deepwater BP Oil Spill," September 17, 2010, http://app.restorethegulf.gov/go/doc/2931/900375/. These numbers reflect a daily snapshot, not cumulative impacts.

[4] Unified Command for the BP Oil Spill, "Statement from Admiral Allen on the Successful Completion of the Relief Well," September 19, 2010, http://app.restorethegulf.gov/go/doc/2931/900707/.

[5] Unified Command for the BP Oil Spill, "Gulf Coast Incident Management Team Established in New Orleans," September 20, 2010, http://app.restorethegulf.gov/go/doc/2931/901235/.

[6] This report will not address questions of liability or compensation. These areas are addressed in other CRS reports, including CRS Report R41262, *Deepwater Horizon Oil Spill: Selected Issues for Congress*, coordinated by Curry L. Hagerty and Jonathan L. Ramseur; also see CRS Report R41308, *The 2010 Oil Spill: Criminal Liability Under Wildlife Laws*, by Kristina Alexander.

[7] The U.S. Gulf coastline consists of the following: Florida (Gulf portion), 770 miles; Alabama, 53 miles; Mississippi, 44 miles; Louisiana, 397 miles; and Texas, 367 miles. See "Coastline of the United States," http://www.infoplease.com/ipa/A0001801.html.

[8] An emergent wetland is one that has plants, such as cattails and bullrushes, whose roots and part of the stem are below water level with the rest of the plant above water.

[9] See http://response.restoration.noaa.gov/book_shelf/827_ERD_ESI.pdf.

[10] Spill data as of Aug. 3, 2010, based on maps from National Oceanic and Atmospheric Administration, available at *http://gomex.erma.noaa.gov/erma.html#x=-90.42000&y=*28.03000&z=6&layers=3796+8440+7303+5723.

[11] Susan-Marie Stedman and Thomas E. Dahl, *Status and Trends of Wetlands in the Coastal Watersheds of the Eastern United States, 1998 to 2004*, National Oceanic and Atmospheric Administration, National Marine Fisheries Service and U.S. Department of the Interior, Fish and Wildlife Service, 2008.

[12] Hurricanes Katrina and Rita produced both gains and losses of wetland acreage. In October 2006, it was estimated that the net effect of the 2005 hurricanes was a loss of nearly 217 square miles (139,000 acres) of land just in Louisiana. This estimate was considered preliminary, because some transformations of land to water are permanent, and others are

transitory or temporary. Confirmation of permanent losses from the hurricanes' storm surges is unlikely before several growing seasons have passed. John A. Barras, *Land Area Changes in Coastal Louisiana After the 2005 Hurricanes: A Series of Three Maps*, U.S. Geological Survey, Open-File Report 2006-1274, October 2006.

[13] U.S. Geological Survey, National Wetlands Research Center, U.S. Department of the Interior, "Without Restoration, Coastal Land Loss to Continue," May 21, 2003, *http://www.nwrc.usus.gov/releases/pr03_004.htm*.

[14] U.S. Geological Survey, Marine and Coastal Geology Program, "Louisiana Coastal Wetlands: A Resource At Risk," November 3, 1995, http://marine.usgs.gov/fact-sheets/LAwetlands/lawetlands.html.

[15] U.S. Geological Survey, "Barrier-Island Inundation and Overwash: Application to the Deepwater Horizon Oil Spill," June 15, 2010, http://www.usgs.gov/deepwater_horizon/scenario/.

[16] An overview of coastal ecology, particularly in the upper Gulf of Mexico and coastal Louisiana, can be found in Michael J. Mac, Paul A. Opler, and Catherine D. Pucket, et al., *Status and Trends of the Nation's Biological Resources, Vol. 1*, U.S. Geological Survey, Reston, VA, 1998, pp. 384-430.

[17] Most bird species use the same routes going north or south. Of those that stop for feeding, the location of the stop would likely be similar in spring and fall. A few species use one route in the spring and a somewhat different route in the fall.

[18] Laura Riley and William Riley, *Guide to the National Wildlife Refuges: How to Get There and What to See and Do* (Garden City, NY: Anchor Press, 1981), p. 238.

[19] See FWS press release of July 8, 2010, "Using Sound Science, Fish and Wildlife Service Addresses Urgent Habitat Needs for Birds and Other Wildlife Along the Gulf Coast," available at http://www.fws.gov/news/NewsReleases/showNews.cfm?newsId=33448A43-ECBE-6890-C4B5296DBEDF3EAC.

[20] More information about the NRCS program may be found at http://www.nrcs.usda.gov/news/nrcs_migratory_birds.html.

[21] 16 U.S.C.A. § 1531.

[22] See http://www.deepwaterhorizonresponse.com/external/content/document/2931/895663/1/Consolidated%20Wildlife%20Table%20091010.pdf.

[23] National Research Council of the National Academies, *Oil in the Sea III, Inputs, Fates, and Effects* (Washington, D.C.: The National Academies Press, 2005), p. 157.

[24] Dennis F. Whigham, Stephen W. Broome, and Curtis J. Richardson, et al., Statement of the Environmental Concerns Committee, Society of Wetland Scientists, "The Deepwater Horizon Disaster and Wetlands," http://www.sws.org/docs/ SWS_OilEffectsOnWetlands.pdf.

[25] Testimony of Jane Lyder, Deputy Assistant Secretary for Fish and Wildlife and Parks, Department of the Interior, before the House Natural Resources Subcommittee on Insular Affairs, Oceans and Wildlife, June 10, 2010.

[26] Holly K. Ober, *Effects of Oil Spills on Marine and Coastal Wildlife*, Dept. of Wildlife Ecology and Conservation, Univ. of Florida, 2010, http://www.wec.ufl.edu/Effects%20of%20oil%20spills%20on%20wildlife.pdf.

[27] See http://www.abcbirds.org/newsandreports/releases/100504.html.

[28] See http://www.evostc.state.ak.us/facts/qanda.cfm.

[29] See FWS Consolidated Fish and Wildlife Collection Report at http://www.fws.gov/home/dhoilspill/pdfs/ collection_06182010.pdf.

[30] Data on bird deaths from oil spills are scarce. The ratio of carcasses to actual deaths could be higher or lower that those in Prince William Sound. In addition, this figure assumes that all birds captured alive survived. To the extent that they did not, the estimates of actual deaths would be increased.

[31] Holly K. Ober, *Effects of Oil Spills on Marine and Coastal Wildlife*, Dept. of Wildlife Ecology and Conservation, Univ. of Florida, 2010, http://www.wec.ufl.edu/Effects%20of%20oil%20spills%20on%20wildlife.pdf.

[32] http://www.fws.gov/home/dhoilspill/collectionreports.html.

[33] See http://gulfseagrant.tamu.edu/oilspill/facts_impacts.htm.

[34] See http://www.noaanews.noaa.gov/stories2010/20100805_hurricaneupdate.html.

[35] National Oceanic and Atmospheric Administration, U.S. Department of Commerce, "NOAA's Oil Spill Response, Hurricanes and the Oil Spill," May 27, 2010, http://www.nhc.noaa.gov/pdf/hurricanes_oil_factsheet.pdf.

[36] For discussion of the Louisiana protective berm project, see CRS Report R41262, *Deepwater Horizon Oil Spill: Selected Issues for Congress*, coordinated by Curry L. Hagerty and Jonathan L. Ramseur.

[37] Adm. Thad Allen Press Conference, June 5, 2010, http://www.deepwaterhorizonresponse.com/go/doc/2931/620563/.

[38] Qianxin Lin, Irving A. Mendelssohn, and Charles B. Henry, Jr., et al., "Effect of Oil Cleanup Methods on Ecological Recovery and Oil Degradation of *Phragmites* Marshes," *1999 International Oil Spill Conference, Paper ID #250*, 1999, http://www.iosc.org/papers/00598.pdf.

[39] As of September 17, approximately 1.84 million gallons of total dispersant have been applied—1.07 million on the surface and 771,000 sub-sea. See: http://app.restorethegulf.gov/go/doc/2931/900375/.

[40] National Research Council of the National Academies, *Oil in the Sea III, Inputs, Fates, and Effects* (Washington, D.C.: The National Academies Press, 2005), p. 144.

[41] Testimony of Aaron Viles, Gulf Restoration Network, before the House Natural Resources Subcommittee on Insular Affairs, Oceans and Wildlife, June 10, 2010.

[42] S.R. Pezeshki, R.D. DeLaune, and A. Jugsujinda, et al., "Major Field Test Evaluates a Shoreline Cleaner to Save Oiled Marsh Grass," *1997 International Oil Spill Conference*, 1997. http://www.iosc.org/papers/01172.pdf.

[43] According to the government, as of mid-September, 411 controlled burns have been conducted and have removed more than 11 million gallons of oil from open waters. See: http://app.restorethegulf.gov/go/doc/2931/900375/.

[44] National Research Council of the National Academies, *Oil in the Sea III, Inputs, Fates, and Effects* (Washington, D.C.: The National Academies Press, 2005), p. 144.

[45] U.S. Environmental Protection Agency, "National Contingency Plan Product Schedule," http://www.epa.gov/ emergencies/content/ncp/product_schedule.htm.

[46] *ScienceDaily*, "Caution Required for Gulf Oil Spill Clean-Up, Bioremediation Expert Says," May 4, 2010, http://www.sciencedaily.com/releases/2010/05/100504142110.htm.

[47] National Research Council of the National Academies, *Oil in the Sea III, Inputs, Fates, and Effects* (Washington, D.C.: The National Academies Press, 2005), pp. 142, 145.

[48] The following section of this report is based in part on CRS Report R41262, *Deepwater Horizon Oil Spill: Selected Issues for Congress*, coordinated by Curry L. Hagerty and Jonathan L. Ramseur.

[49] See EPA "National Contingency Plan Overview" at http://www.epa.gov/emergencies/content/lawsregs/ncpover.htm.

[50] For a list of member agencies, see http://www.nrt.org/production/NRT/NRTWeb.nsf/MADispForm?Openform.

[51] Natural resource damages refer to the environmental impacts caused by an oil spill that result in an actual adverse effect on the resource (e.g., animals, plants, and their habitats) or on the services provided by that resource to the public (e.g., drinking water, recreation). When a spill occurs, natural resource trustees conduct a natural resource damage assessment to determine the extent of injury, and the party responsible for the spill is liable for the damages. In the Deepwater Horizon case, the natural resource trustees include NOAA, FWS, the National Park Service, state trustees from Louisiana, Mississippi, Alabama and

Florida, tribal trustees, and BP. See http://www.response.restoration.noaa.gov/book_shelf/1959_deepwater-Horizon-NRDA-ORR-web-5-7-10.pdf. For more on the Natural Resource Damage Assessment process, see CRS Report R41396, *The 2010 Oil Spill: Natural Resource Damage Assessment Under the Oil Pollution Act*, by Kristina Alexander.

[52] U.S. National Response Team, "Oil Spill Response Strategies for Coastal Marshes during the Deepwater Horizon MC252 Spill," June 15, 2010, http://www.nrt.org/Production/NRT/NRTWeb.nsf/AllAttachmentsByTitle/SA-1061NRT_Marsh_Cleanup_Options_DWH.06032010.pdf/$File/NRT_marsh_cleanup_overview_6-15.pdf?OpenElement.

[53] The teams generally consisted of four persons, representing the federal government (usually NOAA, sometimes FWS), the affected state, and the responsible party. They often were accompanied by a contract archaeologist who could identify historic sites or structures to be avoided during cleanup, such as civil war artifacts or Indian mounds. The responsible party provided logistics (e.g., helicopters and boats).

[54] Douglas Helton, Incident Operations Coordinator, Office of Response and Restoration, National Oceanic and Atmospheric Administration, U.S. Department of Commerce, personal communication, June 24, 2010.

[55] For a graphic description of this process, see http://www.washingtonpost.com/wp-srv/special/nation/one-birds-odyssey/?sid=ST2010061502171.

[56] "Should Oiled Birds Be Cleaned?" by Nell Greenfieldboyce, National Public Radio, June 14, 2010. Available at http://www.npr.org/templates/story/story.php?storyId=127749940.

[57] National Research Council of the National Academies, *Oil in the Sea III, Inputs, Fates, and Effects* (Washington, D.C.: The National Academies Press, 2005), pp. 145-146.

[58] Katherine Unger, "Comparing Deepwater Horizon to Exxon Valdez," The Wildlife Society. Available at http://joomla.wildlife.org/index.php?option=com_content&task =view&id=689&Itemid=321.

[59] For example, see effects of such cleaning at Kenai Fjords National Park, available at http://www.nps.gov/kefj/ naturescience/upload/KEFJ_EVOS_1989-2009_qa.pdf.

[60] Charles H. Peterson, Stanley D. Rice, and Jeffrey W. Short, et al., "Long-term Ecosystem Response to the Exxon Valdez Oil Spill," *Science*, vol. 302 (December 19, 2003), pp. 2082-2086.

[61] Lisa Busch, "Science Under Wraps in Prince William Sound," *Science*, vol. 252 (May 10, 1991), pp. 272-273.

[62] Ibid.

[63] Stephanie Pain, "Exxon Gags Sound Researchers," *New Scientist*, no. 1860 (February 13, 1993).

[64] For example, see Linda Hooper-Bui, "Opinion: The oil's stain on science," *The Scientist*, vol. On-line edition (August 5, 2010). Available at http://www.the-scientist.com/news/print/57610/. Hereafter referred to as "Opinion: the oil's stain on science."

[65] Mark Tran, "BP denies 'buying silence' of oil spill scientists," *The Guardian*, July 23, 2010, http://www.guardian.co.uk/environment/2010/jul/23/bp-oil-spill-scientists-silence/print. For text of letter to Attorney General Eric Holder from the National Wildlife Federation and others, see http://www.nwf.org/News-and-Magazines/Media-Center/Resources-for-the-Press/~/media/PDFs/Wildlife/Oil-Spill-Transparency-Letter.ashx.

[66] Dept. of the Interior. *Confidentiality Agreement, Mississippi Canyon 252 Oil Spill*. Undated. Copy available from author.

[67] See "Opinion: the oil's stain on science" (cited above), Editor's note, citing Pete Tuttle, FWS environmental contaminant specialist, and Dept. of the Interior NRDA coordinator.

[68] Lauren Schenkman, "After Outcry, Oil Data Inches Into the Open," *Science*, vol. 329 (August 20, 2010), pp. 888-889. Available at http://www.sciencemag.org/cgi/reprint/329/5994/888-a.pdf. (Hereafter cited as "After Outcry.")

[69] See http://www.bp.com/genericarticle.do?categoryId=2012968&contentId=7065262, viewed Sept. 30, 2010. Access to BP research funded outside of GRI was not covered in the press release.

[70] See "After Outcry," cited above. CRS was able to find no indications that standards for disclosure of scientific information have been modified for employees in DOI agencies.

[71] Douglas Helton and Tony Penn, "Putting Response and Natural Resource Damage Costs in Perspective," *1999 International Oil Spill Conference*, no. #114, http://www.iosc.org/papers/01767.pdf. The authors examined cost data on 48 U.S. spill incidents between 1984 and 1997 and found that, overall, response costs were the largest category. However, in the case of the *Exxon Valdez* spill, approximately 30% of the total known costs (which exceed $9 billion, according to the authors) were response costs, and 60% were third party claims. Natural resource damage and assessment costs comprised 10% of the known costs. For an overview of natural resource damage assessment laws and policies, see CRS Report R41262, *Deepwater Horizon Oil Spill: Selected Issues for Congress*, coordinated by Curry L. Hagerty and Jonathan L. Ramseur.

[72] U.S. Department of Commerce, National Oceanic and Atmospheric Administration, Office of Response and Restoration, "Has Prince William Sound Recovered From the Spill?," June 2, 2005, http://response.restoration.noaa.gov/topic_subtopic_entry.php?RECORD_KEY%28entry_subtopic_topic%29=entry_id,subtopic_id,topic_id&entry_id%28entry_subtopic_topic%29=258&subtopic_id%28entry_subtopic_topic%29=13&topic_id%28entry_subtopic_topic%29=1.

[73] For FWS employees, for example, see "Service Stories: Tight quarters and a 4:30 a.m. wakeup call on a 'flotel' for oil spill responders," available at http://www.fws.gov/home/dhoilspill/index.html.

[74] See "Oil Spill Response: Who Decides What to Do?", above.

[75] For example, the Coastal Wetlands Planning, Protection and Restoration Act of 1990, also known as the Breaux Act, provided funding to carry out coastal wetland restoration projects in Louisiana and other states. Similarly, the 2006 Gulf of Mexico Energy Security Act provided revenues from offshore oil and gas activities for wetland restoration and other activities in states bordering the Gulf.

[76] For example, in May the federal government granted a portion of Louisiana's request to dredge sand and other material to build berm structures offshore of existing barrier islands in order to reduce inland movement of oil. The state had sought permission to build 128 miles of barrier berms; the government authorized 45 miles of structures. As described in CRS Report R41262, *Deepwater Horizon Oil Spill: Selected Issues for Congress*, coordinated by Curry L. Hagerty and Jonathan L. Ramseur, this proposal has been controversial. For example, in commenting on the state's request, NOAA, EPA and other federal agencies raised a number of concerns, including that the project could deplete finite Louisiana sand resources, thus affecting future high-priority barrier island restoration projects and largely negating two decades' of planning efforts. See "NOAA Considerations on Louisiana Proposal for Barrier island Restoration Plan" in U.S. Army Corps of Engineers, "Corps decision on state's emergency permit request," May 27, 2010, http://155.76.147.200/news/Emergency%20Permit%20Documents%20Compressed%20FINAL.pdf, p. 11.

[77] For information on congressional and legislative activity, see CRS Report R41407, *Deepwater Horizon Oil Spill: Highlighted Actions and Issues* , by Curry L. Hagerty and Jonathan L. Ramseur.

[78] For general discussion of these questions, see CRS Report R41262, *Deepwater Horizon Oil Spill: Selected Issues for Congress*, coordinated by Curry L. Hagerty and Jonathan L. Ramseur.

In: Impacts of the Gulf Oil Spill on Fishing … ISBN: 978-1-61470-971-8
Editors: C. Scarnati and E. Popeo

Chapter 3

REBUILDING AN APPETITE FOR GULF SEAFOOD AFTER DEEPWATER HORIZON*

National Commission on the BP Deepwater Horizon Oil Spill and Offshore Drilling

Staff Working Papers are written by the staff of the National Commission on the BP Deepwater Horizon Oil Spill and Offshore Drilling for the use of members of the Commission. They do not necessarily reflect the views either of the Commission as a whole or of any of its members. In addition, they may be based in part on confidential interviews with government and non-government personnel.

"You can't truth your way out of it..."

"The oil didn't do us any good but it sure didn't hurt Florida fisheries," says Lee Schlesinger, spokesman for the Florida Fish and Wildlife Conservation Commission. "Still, the idea that our seafood may not be safe is stuck in people's minds. People saw the oil gushing every night on CBS and they saw that Florida was on the same map. You can't truth your way out of it."[1] Florida temporarily closed a small portion of state waters to fishing last summer due to the BP oil spill but Schlesinger says the closure had little *direct*

* This is an edited, reformatted and augmented version of National Commission on the BP Deepwater Horizon Oil Spill and Offshore Drilling publication, Staff Working Paper No. 16.

effect on the Florida fishing industry: The closure was brief and did not occur in an area vital to commercial fisheries. Gulf seafood has since been subjected to months of testing pursuant to a seafood safety protocol established by state and federal agencies in response to the spill. So far, no sample has tested positive for contamination near the thresholds considered harmful to human health.*

Schlesinger's comment begs the question: What is the path to effective restoration of consumer confidence in the Gulf brand? If states suffering from a tarnished image cannot "truth their way out of it," what other tools do authorities have at their disposal? And who should bear the cost?

The purpose of this paper is to describe the evolution of the seafood safety debate and its impacts, to discuss ongoing federal and state efforts to diffuse the controversy, and to offer suggestions for avoiding such a crisis of public confidence in the future. Accordingly, the paper is divided into four parts. Part I tracks the development of the post-spill seafood policy from the first federal fishery closures through the implementation of the joint federal-state "Reopening Protocol." Part II describes the impact of the fishery closures on consumer perception and the ramifications for the Gulf economy. Part III highlights the experience of key Gulf States in trying to combat the perception gap through marketing and independent seafood testing efforts. Finally, Part IV presents the lessons learned and suggests that governments can help to rebuild public trust—now and in the future—by promoting greater transparency, citizen involvement, and more robust and independent seafood testing programs.

More than eight months after the Deepwater Horizon oil spill began and five months after the Macondo well was capped, Gulf seafood reaching the consumer market is arguably the safest in the world. The amount of testing by state and federal agencies, the layers of redundancy built into the system, intense public scrutiny and third-party audits, have all conspired to make Gulf seafood one of the most carefully vetted food products on the market today. Well in excess of 10,000 seafood specimens—including shrimp, crabs, oysters and fish—have been collected from affected waters to comprise the thousands of samples that have been tested.** State and federal authorities maintain that

* As subsequently discussed in this paper, however, a coalition of non-profit groups challenge some of the fundamental assumptions of the seafood safety protocol, arguing that the standards employed are not sufficiently protective of human health.

** In correspondence with Commission staff, NOAA officials reported that—in federal waters alone—investigators collected nearly 10,000 animals that comprise nearly 4,600 samples for re-opening purposes and 4,000 samples for baseline and surveillance investigations.

not a single sample has returned from the lab with contaminant levels surpassing the threshold concentrations considered dangerous to human health.[†] For all samples, in fact, the numbers returning from the lab are hundreds to thousands of times below the agreed upon thresholds.[2]

Nevertheless, a substantial portion of the American public remains skeptical, choosing to forego Gulf seafood in favor of "safer" alternatives.[3] A photo taken during the Summer 2010 Taste of Chicago event—more than 750 miles from the Gulf—shows a vendor sign assuring customers that the seafood for sale is not from the Gulf Coast. A national poll by the Associated Press in mid-August 2010 showed that more than half of respondents did not believe Gulf seafood was safe,[4] and a newspaper headline from December declared, "Safety of Gulf seafood debated 8 months after BP oil spill."[5]

This perception gap—between what the science shows and the people believe—is a big problem for Gulf states struggling to repair their bruised economies in the wake of the Deepwater Horizon disaster. Fishing and tourism, along with oil, are the lifeblood of the region, providing tens of thousands of jobs and billions of dollars in revenue each year.[6] Beyond the regional impacts, which can seem abstract and impersonal, are the individual struggles: Despite the fact that more than 99 percent of Gulf fisheries are now open and certain fish populations are soaring,[7] scores of fishermen remain hampered by low sales.[8] Locals spar over the truth of government assertions of seafood safety.[9] Restaurateurs and small business owners, in the Gulf and in surrounding states, continue to face a wary public.[10] Subsistence fishermen, who rely on seafood as part of their daily diet, remain anxious about the health of the waters that nourish them and question whether their catch is safe to eat.[11]

From a technical standpoint, the seafood safety debate derives from the processes of hydrocarbon bioaccumulation and biomagnification.[12] Bioaccumulation is the process by which plants and animals store chemicals—in this case, the building blocks of oil known as hydrocarbons—in their bodies. Biomagnification, in turn, is the process by which increasingly higher concentrations of chemicals are stored in animal tissues as you move up the food chain.

Email from John Oliver, NOAA, January 18, 2010. See section I.B.2.for an explanation of the difference between seafood "specimens" and "samples."

† NOAA officials report that in conducting their analysis of seafood samples, finfish and shrimp, from federal waters, none of the samples had concentrations of polycyclic aromatic hydrocarbons (PAHs) exceeding the levels of concern. For all samples, the concentrations of PAHs were well below the PAH levels of concern.

Vendor Sign at Taste of Chicago 2010. Photo credit: Albert Ettinger.

Marine organisms in the vicinity of the *Deepwater Horizon* spill may have taken up hydrocarbons from the water, food, or sediments. How much oil a given organism potentially accumulates depends on the hydrocarbon concentration in the water, food or sediment the organism encounters, as well as the characteristics of the species, particularly how much lipid (fat) its tissues contain. Hydrocarbons that are not readily metabolized become stored in the organism's tissues. With additional or prolonged exposure, increasing concentrations may be stored. While oil and water do not mix, oil and lipids mix very well, so high concentrations of hydrocarbons may be found in lipid-rich tissues like the liver, brain, kidneys and ovaries. By contrast, lean muscle tissue generally has the lowest concentration of hydrocarbons.

The hydrocarbons of particular concern are polycyclic aromatic hydrocarbons (PAHs), which contain the most potentially toxic components of oil. PAHs can be passed to fish by direct exposure and through various routes in the food chain (e.g., from water to phytoplankton to zooplankton to fish) as smaller fish are eaten by larger ones. However, because animals that are several steps up the food chain, like small fish, have the capacity to metabolize PAHs fairly rapidly, their predators are not likely to accumulate large doses of PAHs from eating them. So while bioaccumulation of toxic oil components does occur in fish, biomagnification does not. This is one reason why the flesh of fish, which may have initially assimilated concentrations of PAHs that would be considered unsafe, may eventually test well below any levels of concern for human consumption. Nevertheless, the distinction between bioaccumulation and biomagnification is nuanced and can be hard for the public to distinguish, particularly in light of the common knowledge that certain kinds of toxins, including heavy metals such as mercury, can biomagnify as you move up the trophic scale. While smaller prey fish cannot readily eliminate mercury metabolically, they can readily metabolize PAHs. It is important to note that the ability to metabolize PAHs quickly only pertains to fish. Other staples of the Gulf seafood industry, including crabs, oysters, shrimp, clams, and scallops, do not as readily metabolize PAHs and may have elevated levels in their fatty tissues.

"Restoration" in the context of the Gulf oil spill generally refers to the need to restore the natural resources harmed by the spill. But the path to effective restoration requires restoring public confidence as well. The persistent controversy over the safety of Gulf seafood—and the associated economic and psychological effects—underscores the compelling need to restore public confidence and the challenges of doing so.

I. Developing a Policy on Seafood Safety

A. The Emergency Closures

1. Federal Waters

By June 2, 2010, thirty-seven percent of federal waters in the Gulf of Mexico had been closed to fishing in response to the *Deepwater Horizon* oil spill. Such a vast closure of federal fisheries was unprecedented; indeed, a federal fishery had been closed only once before as the result of an oil spill.* NOAA's National Marine Fisheries Service initially responded by closing parts of the Gulf exclusive economic zone (EEZ)† to all fishing through two emergency rules effective May 2 and May 7, 2010, respectively.[13] The closures covered 17,624 square miles of federal waters surrounding the Macondo well, or 7.5 percent of the total area of the Gulf EEZ. Each of the emergency rules was set to apply for a ten-day window but—before either rule expired—a new rule was passed that expanded the reach and duration of the former rule. The third emergency rule, published May 14, 2010, provided that all closures would "remain in effect until terminated by subsequent rule-making, which will occur once the existing emergency conditions from the oil spill no longer exist."[14] This third emergency rule made explicit the need to

* The Barge North Cape oil spill off the coast of Rhode Island on January 19, 1996 was the first time a federal fishery was closed due to oil pollution.

† Under the law of the sea, an exclusive economic zone (EEZ) is a sea zone over which a nation has special rights over the exploration and use of marine resources. It stretches from the seaward edge of the nation's territorial sea out to 200 nautical miles from its coast. As applied to the U.S., the EEZ is tantamount to "federal waters."

protect the public from possible exposure to "adulterated" seafood, or seafood that had been contaminated by oil from the Macondo well.

The Oil Pollution Act of 1990 does not itself provide authority to close federal or state waters affected by an oil spill beyond that necessary to enable an effective spill response and clean-up. Rather the management of federal fisheries falls entirely under the purview of the Magnuson-Stevens Fishery Conservation and Management Act (Magnuson-Stevens Act). [15] The Act was originally passed in 1976 to promote the U.S. fishing industry's exploitation of coastal fisheries while simultaneously recognizing that overfishing will undermine that mission. A central role for fishery managers under the Magnuson- Stevens Act is to close fisheries as appropriate to allow depleted fish stocks to replenish themselves. The related provisions of the Act are not applicable to manmade disasters or oil spills, providing closure authority in response to overfishing only. However, the Act separately provides limited authority to close fisheries in the case of an "emergency" under Section 305(c), which provides:

> EMERGENCY ACTIONS AND INTERIM MEASURES.—
> (1) If the Secretary [of the Department of Commerce] finds that an emergency or overfishing exists or that interim measures are needed to reduce overfishing for any fishery, he may promulgate emergency regulations or interim measures necessary to address the emergency or overfishing.[16]

Under authority delegated by the Secretary of Commerce, NOAA invoked Section 305(c) to effect the temporary closures after the Deepwater Horizon well blowout. While this provision allows the agency to close federal waters in the case of an emergency, it does not enumerate any particular criteria for such closures. Thus, NOAA developed its own criteria and made the first emergency closure decisions based on areas of visible surface oiling. Other criteria for closure were developed over time, with input from the Food and Drug Administration (FDA) as well as the five Gulf States.

2. State Waters

In state waters,* marine resources—such as fisheries, oyster beds, and shrimp spawning grounds—belong to the state. If there is a threat of

* State waters typically extend three miles from shore, after which federal waters begin. State waters from certain areas may extend further out to sea, depending on territorial

contamination to such resources, either from a manmade pollution event or natural causes, states retain authority to close affected waters until the danger has passed.[†] This authority is typically delegated to the relevant state health or environmental agency through law or regulation.[‡] Thus, the emergency closure rules promulgated by NOAA immediately after the spill began did not apply to state fisheries. But as new oil continued to gush from the Macondo well, and as existing slicks were pushed toward shore by wind and ocean currents, state authorities had to mobilize also to prevent the harvest of contaminated seafood from state waters.

Exactly one week after the Deepwater Horizon rig explosion, Louisiana announced the closure of select fishing areas and oyster harvesting beds as a precautionary response to the oil spill. [17] This first rule included a significant closure of state oyster grounds in the coastal parishes of Plaquemines and St. Bernard.[18] As the Gulf state nearest the Macondo well, and the one with the most extensive coastal waters —over eight million acres—Louisiana issued a spate of additional closings in May, June and July. In Louisiana, the Department of Wildlife and Fisheries and the Department of Health and Hospitals have primary control over the management of state marine resources: The former has jurisdiction over state fisheries, while the latter has specific authority over the state's oyster harvesting areas due to the human health risks posed by consuming raw oysters.[§] These agencies already conducted periodic sampling and testing as a matter of course before the spill, but the frequency of testing increased dramatically after the spill. In April and May, these agencies collected and tested samples from state fisheries and oyster beds to establish a baseline for comparison to samples collected after oiling.[**]

---

arrangements. For example, the Florida fisheries closure in Escambia County extended out nine nautical miles from shore.

[†] In the case of an oil or hazardous substance spill, the U.S. Coast Guard retains primary authority over cleanup and response and can close off areas of state waters as needed to facilitate response. This type of closure is, however, distinct from a closure specifically applicable to marine resources—a fishery or oyster reef, for example—which is entirely the purview of the state environmental agency.

[‡] In Florida, for example, the state Constitution delegates authority for fisheries closures to the Florida Fish and Wildlife Conservation Commission.

[§] Oyster contamination is most often due to the presence of coliform bacteria, from human and animal waste. Such pathogens are generally killed off in the cooking process, but because oysters are consumed raw, there is a need for heightened surveillance by state health authorities.

[**] Such baseline tests can establish the threshold levels of contaminants for various marine organisms in the area. It is unclear to what extent Louisiana was able to complete such testing before oil hit and what a comparison to post-spill samples showed.

On May 3, the Mississippi Department of Environmental Quality and the Mississippi Department of Marine Resources announced they were "continuing comprehensive preparations," including the placing of protective boom around ecologically sensitive estuaries and marshes,[††] as well as baseline testing at multiple sites to document pre-spill conditions of coastal ecology.[19] In an early effort to combat negative public perception, the state issued a press release on May 7, emphasizing that all state waters were open to fishing and that "Mississippi seafood is safe to consume."[20] Nevertheless, Mississippi was forced to close some state waters on June 1 due to the presence of oil in state waters.[21] State authorities made eight additional closures decisions in June. On July 1, the state closed the Mississippi Sound to all recreational and commercial fishing activities.

On June 14, the Florida Fish and Wildlife Conservation Commission closed a relatively small portion of state waters—about 230 square miles—to the harvest of fish, shrimp and crabs, noting that it was "taking precautionary actions regarding harvest and consumption of these marine species, which may be affected by oil from the BP Deepwater Horizon spill." The closure area covered approximately 23 miles of coastline in Escambia County, where oil from the Deepwater Horizon spill was then present, and out nine nautical miles (10.36 miles) into the Gulf from the Alabama line east to Pensacola Beach.[22]

The Alabama Department of Conservation and Natural Resources closed a portion of state waters to all fishing activity from the east end of Dauphin Island to the Mississippi state line on June 2.[23] Additional waters were closed on June 10, including Mississippi Sound, Heron Bay, Portersville Bay and Grand Bay.[24] Mobile Bay remained open to fishing for the duration of the spill, except for a small area at the mouth of the bay.

With the possible exception of tar balls and very light oil sheens, oil from the Macondo well never made landfall in Texas and no state fisheries were closed. An "Emergency Notice" was posted on Texas Parks and Wildlife Department website on July 28, noting that "[Department] biologists are aware of the spill and its potential impact on fishing in Texas waters, and will work with partner agencies to take action if needed to manage fisheries and safeguard public health."[25]

[††] Coastal estuaries and marshes are the natural nurseries for many kinds of marine life including shrimp, crabs, oysters, and fish

B. State and Federal Collaboration

While states are autonomous over their sovereign waters, the unprecedented scope of the *Deepwater Horizon* oil spill militated in favor of a joint approach. As mentioned above, Louisiana began closing its state waters to fishing just days after the spill began, according to its own criteria, before oil ever reached state waters. Other Gulf States followed suit.[26] It quickly became clear that a coordinated effort was needed to avoid a patchwork of disparate state closure rules after the spill.

In May 2010, a working group was formed, including public health and environmental agency officials from the five Gulf States, NOAA, the Food and Drug Administration (FDA) and the Environmental Protection Agency (EPA), to devise a uniform protocol to guide Gulf fishery closures and reopening decisions in federal and state waters. Through a series of meetings and conference calls over the next weeks and months, the working group hammered out protocols for fishery closures, seafood testing, and fishery reopen ings.[27]

State and federal officials have since used a multi-pronged approach to try and ensure seafood harvested from Gulf waters is not contaminated by oil. The strategy includes precautionary closures, market surveillance and a reopening protocol.[28]

1. Precautionary Closures

On June 14, 2010, NOAA and FDA—with input from the Gulf States—published a "Fishery Area Closure and Surveillance Plan" that required the following areas to be closed:

- Areas where there is any visible oil on the surface;
- Areas where there is clearly detectable levels of subsurface oil from the BP oil spill;
- Areas that do not currently have surface oil but where NOAA projects there will be surface oil based on its 48-72 hour surface oil trajectory forecasts and subsurface oil locations;
- A five mile buffer surrounding any closed area as a precautionary measure to account for "uncertainties in the actual boundaries of the oil and movement of fish."[29]

These remain the operative criteria for closure in federal and state waters today, allowing NOAA and state authorities to close additional areas as needed. As described below, NOAA was forced to make an additional closure as late as November 2010, when shrimpers pulled up subsurface oil in their nets.

2. Dockside Sampling and Market Surveillance

The Fishery Area Closure and Surveillance Plan also set forth the parameters of a new working relationship between NOAA and the FDA. As the lead federal agency on food safety, FDA's input was critical to developing the standards for seafood testing that would be used as a basis for reopening closed fisheries. The plan further provided that NOAA would take the lead on ocean surveillance and dockside sampling, while the FDA would take steps to ensure seafood safety further along the supply chain.

As a first line of defense, NOAA uses vessel tracking systems and information from on-board observers to verify that fishermen are not operating in closed areas.[30] NOAA conducts limited dockside sampling as a further check—collecting, at random, seafood samples at the dock for laboratory analysis. FDA provides another layer of protection by testing samples further along the supply chain. During the summer of 2010, for example, the FDA collected and tested samples from each of the roughly 65 primary seafood processors handling Gulf seafood at the time.[31*] A primary processor buys seafood directly from the fisherman, so surveillance at this stage is especially crucial. FDA also helps seafood processors review their own individual source controls to ensure proper documentation and to prevent seafood obtained from unknown sources from entering commerce. To date, neither NOAA's dockside sampling nor FDA's market surveillance programs have turned up any evidence of contaminated seafood due to the BP oil spill.[32]

3. The Reopening Protocol

The purpose of the Fishery Area Closure and Surveillance Plan was to ensure that contaminated seafood did not make it to market and that closure areas could be expanded as necessary to keep consumers safe. At the same time, there was significant pressure on government to keep commerce moving and to reopen closed fisheries as soon as practicable. To help guide this

* There are roughly 300 primary processors working along the Gulf Coast, but many of these temporarily closed their doors or were securing their seafood from non-Gulf sources during the height of the spill. As of January 2011, the FDA was in the process of conducting a second round of sampling from Gulf seafood processors; agency representatives expected a significantly larger sample pool in light of the fact that many of the processers had returned to work.

determination, the interagency working group also developed a "Reopening Protocol," which was published on June 18, 2010.[33] The Protocol—applicable to both federal and state waters—sought to strike a balance between protecting consumers and unnecessarily crippling the seafood industry. It sets forth the following background principles:

- NOAA and the FDA are working with other federal and state agencies to protect consumers from adulterated and unsafe seafood, while minimizing undue economic burden on any impacted seafood industries.
- Once oil or chemical contaminants are visually observed on the ocean surface, it is recommended that the fishery be closed until free of sheen, and subsequent testing has been completed to confirm that seafood from affected areas are wholesome and safe for human consumption and use in animal feed.
- After the initial fishery closure, the best approach for determining the safety and acceptability of seafood from oil-contaminated areas is one that involves organoleptic analysis of products (i.e. sensory testing) followed by chemical analysis.
- Fishery closure areas also include areas that NOAA predicts will have surface oil as well as, a precautionary buffer zone around known contaminated waters to account for uncertainty. After confirming through subsequent evaluation that oil did not enter an area, the area may be reopened without subjecting seafood samples to evaluation under this protocol. This protocol is an added layer of protection being applied to seafood only in areas known to have been contaminated.*

Beyond this prefatory guidance, the Protocol lists four specific criteria that must be satisfied before a closed area can be reopened:

1. *Low threat of exposure:* The area must be free of oil and oil sheen on the surface;

---

* The fact that seafood from only certain areas of the Gulf is subject to sampling under the Protocol was a source of concern for some local fishermen, who worried about liability issues should consumers become sick after buying their catch. Interviews with Commission staff, August 2010. Several fishermen pointed to the need for more extensive dockside sampling and certification procedures, and suggested that if the government has made the determination that certain fishing grounds are safe and should be reopened, then the burden should be on NOAA/FDA to give its "stamp of approval" to Gulf seafood so that any potential liability can be shared.

2. *Evaluation of oil movement:* There is low risk or threat of future exposure to oil based on the current predictions;
3. *Sensory testing of seafood:* Seafood must pass sensory sampling (small and taste) for oil exposure (taint);[†]
4. *Chemical analysis of seafood:* Sensory testing is just the first step in the analysis: Any sample that passes sensory testing must also pass chemical analysis for oil concentration (toxicity).[34]

Seafood testing is central to the reopening analyses, and the part of the Protocol that has attracted considerable public attention. According to the Protocol, oil exposure presents two kinds of risks to consumers. The first is the presence of petroleum "taint" (a smell or taste of oil), which renders seafood unfit for human consumption. The second concern is the presence of PAHs, which, as described in the primer on bioaccumulation, can pose a chemical hazard to humans at certain concentrations.

The nuts and bolts of the seafood testing process are as follows: Seafood specimens are collected from the area being considered for reopening, and partitioned into samples for testing. Seafood first undergoes sensory testing, during which a minimum of ten experts trained to detect oil- tainted fish assess each sample.[‡] At least seventy percent of the assessors must find no detectable odor from the raw or cooked sample and no detectable taste from the cooked sample. If any sample fails sensory testing, then the entire site slated for reopening fails and the area remains closed. If a sample passes sensory analysis, it must then undergo a more rigorous laboratory chemical analysis. The purpose of the chemical analysis is to measure the concentration of PAHs in the sample, and to ensure it did not exceed consumption thresholds deemed to pose an unacceptable risk to human health.

The absolute value of the consumption thresholds (e.g., 2 parts per million of PAHs in a daily portion) is pre-set according to standard risk assessment methods that require authorities to make assumptions concerning how much

[†] "Even at levels too low to affect human health, PAHs and other chemicals in oil can cause seafood to have an unusual smell or taste. This is called "taint." Taint does not necessarily mean that fish or shellfish are unsafe to eat, but tainted seafood is not allowed to be sold in interstate commerce." NOAA, Office of Response and Restoration, Oil Contaminants in Seafood: *Evaluating the Risk for People Who Eat Fish and Shellfish*, June 1, 2010.

[‡] This is the so-called "sniff test" which was widely ridiculed by critics. Nevertheless, unilateral focus on the sniff test is disingenuous because the Reopening Protocol required BOTH sensory evaluation AND chemical testing—not just one without the other.

seafood consumers eat in order to give meaning to the absolute value.* Thus, the protocol necessarily incorporates a "fish consumption assumption" which guides the application of the risk thresholds. In the case of the Gulf oil spill, authorities agreed to assume that individuals eat no more than two six-ounce portions of fish per week and one three-ounce portion of shrimp (equal to about four jumbo shrimp) per week.†

To address the use of dispersants and how those might affect the safety of seafood, the Reopening Protocol—over the summer and into the fall—stated that:

> To date, available information indicates that the dispersants being used to combat the oil spill do not appear to accumulate in seafood and therefore, there is likely little public health concern from them due to seafood consumption. However, as per this protocol, sensory testing and further work to identify component compounds in known exposed fish will be conducted for dispersants.[35]

This disclaimer was cold comfort to community members convinced that dispersants pose a significant health danger, and who further doubted the value of sensory testing methods—aka "the sniff test."‡

On October 29, 2010, NOAA and FDA announced the addition of a new chemical test for dispersants as a condition precedent to reopening closed

---

* Scientists use a process called "risk assessment" to determine levels of chemical exposure that may cause health problems. For chemicals known to pose a threat to human health, scientists establish "safe" levels of exposure, taking into account sensitive populations such as children, the elderly, and people with weakened immune systems. Exposures below these set levels—or risk thresholds—are considered safe, while exposures above the risk thresholds should be avoided. With respect to PAHs, which are suspected to cause cancer, scientists have set an acceptable cancer risk level, which is extremely low as compared to cancer risks associated with lifestyle choices, such as smoking or poor diet. NOAA, Office of Response and Restoration, *Oil Contaminants in Seafood: Evaluating the Risk for People Who Eat Fish and Shellfish*, June 1, 2010.

† Each of NOAA's reopening decisions was accompanied by a memorandum entitled "Metabolism of PAHs by Teleost Fish" which highlighted the fact that teleost fish (fish with backbones such as grouper, snapper and tuna) are able to rapidly metabolize PAHs such that that these compounds do not become stored in their tissue in any significant quantity. The memorandum stated that even when fish are exposed to oil, there is little chance for PAHs to travel up the food chain to human consumers. Notably, similar claims were not made with respect to other, non-teleost, staples of the Gulf seafood industry, including crabs, oysters, shrimp, clams, and scallops.

‡ Public comments at the Commission's Forum in New Orleans on January 12, 2011 revealed continuing apprehension and anecdotal observations that dispersants were making people sick. See video of New Orleans Forum, January 12, 2010,_http://www.oilspillcommission.gov/forum.

waters.[36] The new test targets one of the main chemicals building-blocks of the dispersants used in the *Deepwater Horizon* oil spill, a compound called dioctyl sodium sulfosuccinate (DOSS). The test is to be used in any future reopening decision, and is applicable to fish, oysters, crab and shrimp. The level of concern is 100 parts per million for fish and 500 parts per million for shrimp, crab and oysters, but official sources suggest the test is able to detect DOSS at levels of 2000 times below the lowest level of concern.[37] Past research suggests that finfish metabolize DOSS rapidly, and subsequent research at NOAA and FDA laboratories in Galveston, Texas and Dauphin Island, Alabama, respectively, provides further support that fish, crustaceans and shellfish quickly clear DOSS from their tissues.[38]

When the new test was announced, NOAA and FDA had already used it to analyze 1,735 samples collected from June to September over a range of locations in the Gulf (including state and federal waters). The samples spanned a range of species, including grouper, tuna, wahoo, swordfish, gray snapper, butterfish, red drum, croaker, and shrimp, crabs and oysters. Of the 1,735 samples tested, 13 showed trace amounts of dispersants residue but they were well below the aforementioned safety thresholds for DOSS. As such, it was concluded that they did not pose a threat to human health. According to FDA Commissioner Margaret Hamburg, "The overwhelming majority of the seafood tested shows no detectable residue, and not one of the samples shows a residue level that would be harmful for humans. There is no question Gulf seafood coming to market is safe from oil or dispersant residue."[39]

In December 2010, FDA also began revamping its testing protocols to incorporate a fluorescence test requirement.* Fluorescence testing is more sensitive than traditional chemical analysis for PAHs. Additionally, authorities are considering incorporating additional tests for different contaminants found in oil aside from PAHs to the testing protocols. For instance, environmental and health NGOs have suggested that a test for heavy metals, such as mercury and arsenic, ought to be included.[40]

* The FDA developed a high-performance liquid chromatography (HPLC) method with fluorescence detection to measure PAH levels in samples from state waters. Such a test is evidently more sensitive than the gas chromatography-mass spectrometry (GC-MS) method formerly in use. NOAA Marine Fisheries has also established this HPLC method at its laboratories in Pascagoula and Seattle, and it is being used in addition to the original GC-MS method. Email from John Oliver, NOAA, January 18, 2010.

C. Community Concerns with Seafood Testing

The seafood testing components of the Reopening Protocol are under scrutiny by environmental and public health groups who claim that the chosen consumption risk thresholds are not sufficiently protective of vulnerable populations, such as subsistence fishermen, pregnant women and children.[41] Additionally, critics of the government protocols take aim at the underlying seafood sampling procedures—where seafood samples were collected and how many—as being opaque and insufficient. Joe Jewell, Deputy Director at the Office of Marine Fisheries at the Mississippi Department of Marine Resources, contends that these concerns stem from a misunderstanding of the process rather than any flaw in design.[42] Nevertheless, Jewell recognizes that authorities may have fallen short in communicating effectively with the public and clarifying misconceptions as they arose.[43]

1. PAH Risk Threshold Issue

The heart of the seafood testing procedure is a comparison of the PAH levels detected in samples to determined "risk thresholds"—namely the maximum concentration of PAHs authorities calculated humans could consume safely, any concentration above which is deemed unsafe.[44] These risk thresholds are based on assumptions regarding how much seafood people in the Gulf eat and how often. To recap, the thresholds assume average consumption of a 6 oz. piece of fish twice a week and a 3-oz serving of shrimp once a week, equal to about four jumbo shrimp. Some Gulf residents take issue with these consumptions assumptions, pointing out, for example, that "four shrimp don't even make a Po' boy."[45]

On October 1, 2010, a coalition of non-government organizations (NGOs) including the Natural Resources Defense Council, the Sierra Club, the Deep South Center for Environmental Justice, and many others sent a letter to the FDA asking it to adjust the PAH risk thresholds applied in the Gulf oil spill to better reflect the higher rates of seafood consumption among Gulf residents, particularly subsistence fishing communities whose consumption rates can be an order of magnitude higher than the national average.[46] The groups maintain that the risk thresholds in use are based on surveys of national seafood consumption rates and, given the region's voracious appetite for seafood, are thus a poor barometer for the Gulf.* The letter suggests that the scientifically-validated consumption rates adopted by the EPA and the World Health Organization for high-volume seafood consumers are more appropriate and should be adopted.

The coalition followed-up with another letter on December 8, [47] which included the results of an informal survey of seafood consumption patterns among Gulf Coast residents conducted between August and October 2010.† The survey of 547 coastal residents in Louisiana, Mississippi, Alabama and Florida, found that many or most respondents reported consuming more fish, shrimp, oysters and crab than FDA risk thresholds presumed.[48] For example, for fish, the survey found that the average, median and 90th percentile consumption frequency of the study population was three, two and five times per week, respectively. By contrast, the FDA assumes residents consume only one meal of fish per week. Likewise, for shrimp, the consumption frequency among respondents was two, two and five times per week, respectively—substantially more than the FDA's assumption of one serving per week. Some subpopulations, moreover, had significantly higher consumption rates. The Asian/Pacific Islander ethnic group had a median shrimp consumption frequency of four times per week and a 90th percentile of six.

While not statistically rigorous, the survey suggests that Gulf residents may eat significantly more seafood than the risk thresholds assume, and highlights the vitality of this debate. As subsequently described, however, all of the laboratory tests conducted to date—for both federal and state waters—have shown PAH levels orders of magnitude below applicable risk thresholds. Thus, for this spill, even if the seafood consumption assumption thresholds were to be increased dramatically (e.g., from two meals per week to 20), potential PAH exposure still would not be judged harmful to human health. Nevertheless, this is not to say that the fish consumption assumption threshold would not be relevant in the case of a future oil spill, where concentrations of PAH's in tested samples may be much higher; every spill is different. Little would be lost if FDA were to commission a survey of seafood consumption

* According to the NRDC, FDA's argument for using national numbers for calculating exposure thresholds in the Gulf was the absence of a reputable survey of seafood consumption patterns in the Gulf.

† It is important to note that the survey was not based on a random sample. Rather, the survey was made available online and paper copies were distributed by community representatives in English and Vietnamese. Thus, respondents were largely self-selected and the reliability and validity of the survey is compromised as a result.

patterns for the Gulf Coast in order to bolster public trust and put to rest this ongoing controversy.[‡]

Still, officials close to the process suggest that the standards in place are already sufficiently protective of Gulf populations. In the first instance, they point out that the applied seafood consumption assumption is the product of the collective best professional judgment of experts from the FDA, NOAA and the Gulf States. A common misconception is that the FDA simply applied a value based on the average national seafood consumption rates, but state officials make clear that they were involved in calculating the risk thresholds that would apply in the wake of the Deepwater Horizon: While it is true that the Reopening Protocol incorporates seafood consumption rates that are based on a national survey, the working group agreed to use the rates corresponding to the 90th percentile of that survey in order to better reflect Gulf realities.[49] In other words, 89 percent of the nation eats less seafood, on average, than the value relied on by the Gulf seafood safety protocol, while 9 percent of the nation eats more.

2. Seafood Sampling Issue

Another oft-repeated criticism of the seafood testing protocol has to do with the sampling procedures used to collect seafood from affected waters. Each federal reopening decision is accompanied by a map identifying sampling locations as well as data reflecting the number and type of specimens collected. Critics portray the sampling process as haphazard and insufficient, leading to premature reopenings of fisheries before the safety of the catch is conclusively established. By way of example, NOAA's website indicates that for the 5,000+ square mile reopening of federal waters which took place on September 2, only 12 shrimp samples were tested, totaling 73 shrimp specimens.[50] While the subsequent analyses of these samples revealed no contamination at levels considered risky to human health, advocacy groups have questioned the paucity of the number of samples analyzed relative to such a large area.[*]

‡ It bears noting that—in light of current chemical test results—even if FDA and NOAA were to substantially increase the seafood consumption assumption, the detectable levels of PAH and dispersants in seafood harvested from the Gulf likely would remain below applicable risk thresholds.

* For example, Dr. Bill Walton, Professor of Marine Fisheries at Auburn University in Alabama, said "that the most recent federal opening was based on what sounds to me like a relatively small sample size, especially when those samples are spread across multiple species. . . . I believe that 1) the number of samples and replicates should be increased to address public concern and 2) the sampling, replication and results should be readily found by the public."

Similar criticisms have been leveled against sampling efforts in state waters, which also follow the federal regulations and protocols. In examining the sampling history for each of the Gulf States, the NRDC found that—by September 2—Louisiana had tested 53 shrimp, Alabama and Florida had each tested 42 shrimp, and Mississippi had tested 21 shrimp. Gina Solomon of NRDC explains why these numbers may be insufficient:

> Even in Louisiana, where more testing has been done, there are still questions: A total of 53 shrimp from Louisiana waters have been chemically analyzed and the levels reported as low. If those 53 shrimp really represent the full extent of the open territory, maybe that's OK, but this oil spill has lots of variability, with pockets of contamination appearing in many locations. As a scientist, I would feel better if I knew there was a robust sampling strategy to address this variability and to take enough samples - maybe several hundred - to assure that if there were contaminated shrimp out there, they would be detected.[51]

Concerns such as these are plausible, and are shared by a number of community groups and NGOs in the region. But state and federal officials suggest there are a variety of factors underlying the sampling protocols which were not well understood by the public, leading to confusion and propagation of misinformation. For example, Mississippi officials point out that the number of samples necessary to justify reopening depends on the location and extent of oiling: While the number of samples taken in Mississippi was low as compared to Louisiana, that is because Mississippi has significantly less shoreline—and state waters are correspondingly smaller in area.[52] Moreover, the extent of oiling in Mississippi state waters was relatively low. NOAA officials confirmed and elaborated on this point, noting that:

> In general, seafood was collected more intensely in areas that were more heavily oiled, for example, closer to the *Deepwater Horizon* wellhead. In addition, finfish and shellfish are not randomly distributed in the Gulf of Mexico; therefore, sampling of particular species was more intense in locations where those species naturally occur. The species collected were also representative of the species targeted by commercial and recreational fishermen in that area.[53]

Written Response of Dr. Bill Walton to Commission Staff Questionnaire on Fisheries, September 5, 2010.

Commission staff also asked NOAA to comment on the criteria that guided the agency in deciding how many specimens to collect. In a written response, NOAA officials stated: "Sample numbers were based on statistical considerations derived from previous efforts in surveys of fish in the Gulf of Mexico, and variation in data in earlier studies using sensory and chemical testing of seafood."[54] Such an explanation is hardly likely to convince skeptics demanding specific information on sampling to allay their concerns.

Another source of confusion cited by officials is the difference between a specimen and a sample. [55] A specimen is what is collected from the sea—one shrimp for example—whereas a sample is what is subjected to sensory or chemical analysis. A specimen and a sample may not be the same. In order for a laboratory to conduct a chemical analysis on biological tissue, a certain volume of tissue is required in order to return confident results. With respect to certain small species, several specimens are needed to comprise a sample for analysis. Unfortunately, the distinction between specimens and samples is not always made explicit and often used interchangeably by authorities when speaking to the public. This lack of necessary precision causes confusion when state and federal authorities proclaim that thousands of samples/specimens were taken, while a significantly lower number is posted on official reports.[56]

Without access to official sampling rationales and methodology, the concerns of groups like NRDC are understandable. The sampling controversy is another example for greater care to public relations and communications would be a win-win situation for all concerned.

D. Reopening the Fisheries

By June 2, 2010, 37 percent of the federal waters in the Gulf of Mexico were closed to fishing due to the *Deepwater Horizon* oil spill.[57] Federal authorities announced the first reopening of federal waters on June 4, when approximately 12 percent of the closure area was reopened. Additional reopening decisions were issued throughout the summer and fall. As of this report, the most recent reopening took place on November 15, when NOAA opened an 8,403 square mile area that had been closed. By that time, 99.6 percent of federal waters were open to fishing.[58]

Each of the reopening decisions has been posted on NOAA's website[59] with extensive supplementary material, including: (1) seafood sample locations, (2) seafood testing data, (3) NOAA's letter to FDA providing the rationale for reopening (i.e., test results showing no contamination), (4) the associated FDA concurrence letter, as well as a copy of (5) the reopening protocol and (6) a background paper regarding the metabolism of PAHs by teleost fish. While the number of specimens collected and sample locations were posted for each reopening, there was no description of the underlying standards or methodology used to inform the sampling decisions. Publication of such a sampling protocol would have gone a long way in mollifying some of the concerns raised by groups such as NRDC concerned with the robustness of the seafood testing regime.

Louisiana began reopening its waters to fishing in early July. By mid-November 2010, over 98 percent of state waters were open to commercial fishing. [60] The most recent reopening decision (as of the printing of this report) on November 8th, applied to crab fisheries, which were historically among the last fisheries to reopen—second only to oyster grounds—due to their heightened sensitivity to oiling impacts.[61] Such a species-specific reopening process was common across affected Gulf States, with waters typically opening to fin-fishing first, and subsequently to fishing for shrimp, crab and oysters.

In Mississippi, reopening also proceeded in stages, with one portion of the closure area reopening to the harvest of finfish and shrimp—but not crab—on July 30 and subsequent reopenings across all state fisheries in August.[62] State officials recoil at allegations that the independence of their seafood testing has been in some ways compromised or that its science is unsound. Officials describe multiple layers of testing: In addition to sending samples directly to NOAA for testing per the joint protocol, the state also sends samples to the State Chemical Lab, where they undergo a separate analysis using even more stringent standards than required by the NOAA protocol. The state inspection and monitoring process is also robust, with inspections beginning dockside and continuing almost to the grocer's shelves.

In Florida, there had been just one closure along the coast of Escambia County. Reopening of the area proceeded in three tiers, according to the marine species at issue: On July 31, the Florida Fish and Wildlife Conservation Commission reopened the area to the harvest of fish; on August 16, the area was reopened for shrimping; and on September 16, the area was opened to the harvest of crabs. With each reopening, state authorities stressed the extent of surface monitoring, seafood testing and analysis that were required before harvest of a given species would be permitted again.

The first reopenings of Alabama waters for fishing took place on August 7,[63] and all remaining Alabama waters were open to fin-fishing on August 16.[64] The August 16 reopening notice highlighted the latest round of test results—including results from new florescence testing methods. The reopening applied to fin-fishing only and subsequent notices provided for the reopening of closed waters to crab harvesting and shrimping. Like Louisiana, Alabama provided a hotline number for citizens to call should they suspect oil contamination.[65]

Seafood safety does not seem to have been a significant issue in Texas and there is no indication that the state is currently considering a request to BP for marketing funds. Calls from Commission staff to the listed media contact for the Texas Parks and Wildlife Commission have not been returned.

E. New Closure in Federal Waters

On November 24, 2010, new reports of subsurface oil forced NOAA to shut down 4,213 square miles of federal waters off Louisiana, Mississippi, and Alabama to fishing for royal red shrimp.* The closure came just nine days after NOAA announced that over 99 percent of previously closed Gulf waters were oil-free and open for business.[66] According to a NOAA press release, the closure decision was made "out of an abundance of caution . . . after a commercial shrimper, having hauled in his catch of the deep water shrimp, discovered tar balls in his net."[67] The tar balls were sent to the U.S. Coast Guard for "fingerprint" analysis to determine whether they contained oil from the Macondo well. NOAA officials have since reported that testing revealed that some of the tar balls were not related to the spill, while the testing of others was inconclusive.[68]

The royal red shrimping ban was lifted on February 2, 2011, after NOAA announced that "shrimp and finfish from the area are safe to eat,"[69] and that all seafood samples passed both sensory and chemical analysis.* As of February 2, an area immediately surrounding the Macondo wellhead— measuring 1,041 square miles—remains closed to fishing. All commercial and recreational fishing, including catch and release, is prohibited in this closed area.[70]

* The new closure was only applicable to royal red shrimping, which involves dragging trawl nets along the ocean floor at depths of over 600 feet. Fishing in shallower waters in the same area was still permitted. The impact of the closure was further limited because only about 250 fishermen currently hold permits to harvest this species.

* The test results are available at *http://sero.nmfs.noaa.gov/deepwater_horizon_oilspill.htm*.

II. Impacts of the Fishery Closures

The American people were obviously relieved when the Macondo well was finally capped on July 15. The event marked the end of the acute phase of the *Deepwater Horizon* disaster, a three-month state of emergency which paralyzed industry and put tens of thousands of Gulf residents in limbo. But a lengthy period of restoration—of damaged ecosystems, battered businesses and faltering public confidence—was just beginning.

A. Persistent Doubts about Seafood Safety

A month after the well was capped and just as the Gulf white shrimp season was set to begin, a national poll conducted by Associated Press-GfK from August 11-16 found that 54 percent of the 1,007 respondents lacked confidence in the safety of Gulf seafood from areas affected by the spill, and 55 percent said they weren't confident that the beaches in affected areas were safe for swimming.[71] The results of a consumer perception survey conducted by the state of Louisiana in August were also sobering.[72] Of the 1,003 nationwide respondents, 53 percent believed, or were not sure whether, Gulf shrimp prices were up because they had to be cleaned of oil; 44 percent believed, or were not sure whether, fishing was allowed where oil is present; and 35 percent believed, or were not sure whether, regulations were in place to ensure that contaminated seafood is not sold. According to Lieutenant Governor Scott Angelle,[73] these weren't even the most serious issues: "The most devastating finding is that 48 percent still believe restaurants that serve Louisiana seafood put customers at risk."[†] Also troubling was the finding that consumer doubts on seafood safety were not improving over time: An identical survey conducted three months earlier found that roughly the same proportions of respondents reported the concerns above.[74] The report authors note that the study "confirms that despite passage of time, the nation's perception has not improved."[75]

National surveys by the University of Minnesota Food Industry Center corroborate that the consumer confidence problem is stubborn and resistant to change.[76] Of respondents surveyed in May 2010, 53.1 percent said they were

† The figure cited by Lt. Governor Angelle is not entirely accurate in light of the fact that it combines the percentage of respondents who said that restaurants serving Louisiana seafood put their customers at risk with the percentage of respondents who said they were "not sure" whether such restaurants put their customers at risk.

eating less seafood based on the Gulf spill, with 24.2 percent saying they would only eat seafood that they knew did not come from the Gulf of Mexico. The percentage was virtually unchanged in subsequent surveys conducted in June, July and August with more than 50 percent of respondents each month stating that their seafood consumption patterns had changed due to the spill. In each of those three months, moreover, more than 21 percent of respondents said they would only eat seafood that did not come from the Gulf.

More recent polls conducted in September and October suggest minimal improvement in consumer confidence: In the September survey, 45.2 percent of respondents said they were eating less seafood due to the Gulf spill, with 17.5 stating categorically that they would not eat seafood from the Gulf. In November, the percentages increased slightly, with 48.3 percent of respondents reporting eating less seafood due to the spill and 18.4 stating they would not eat Gulf seafood. Thus, more than six months after the spill, nearly half of all respondents reported changing their seafood consumption patterns as a direct consequence of the *Deepwater Horizon* oil spill. Most recently, a national survey commissioned by Greater New Orleans, Inc. conducted in November showed slightly more favorable results—at least with respect to consumer perceptions of Louisiana seafood: Of the 1,007 registered voters polled across seven major U.S. media markets, 17 percent reported an unfavorable impression of Louisiana seafood while 58 percent had a favorable impression. Louisiana may have a head start in the recovery of consumer confidence due to their prior state-specific seafood branding and marketing efforts. The survey found that respondents were more favorable to the term "Louisiana seafood" (with 53 percent being likely to purchase) than "Gulf seafood" (with 53 percent being unlikely to purchase).[77]

B. Loss of Public Trust

At least regionally, declines in consumer confidence have been fueled, in part, by an eroding relationship between community groups and local government. During the summer, for example, a group of fishermen attracted public attention with persistent claims that they were discovering oil in reopened areas. None of these claims was ever publicly verified and many were puzzled by what they perceived to be fishermen speaking out against their interests. Passions ran high on both sides. In the most extreme cases, fishermen and their supporters cast themselves as whistleblowers working to uncover what they alleged was a vast government conspiracy. The government in turn painted the fishermen as being less interested in returning to honest work and more interested in continuing to collect BP money.

During the height of the response, when the Vessels of Opportunity Program was in full swing, it was common to hear reports from fishermen that they had discovered tar balls or oil in areas that had been reopened to fishing. In response to such reports, a NOAA Sea Grant agent said that "Skimming oil is a lot easier than trawling for shrimp."[78] The insinuation being that certain fishermen had an incentive to perpetuate concerns about the safety of the fishing grounds in order to extend their participation in the lucrative Vessels of Opportunity Program. Whether any fishermen actually had such ulterior motives or whether they were only voicing good-faith concerns may never be clear, but debates such as this one did create division and impact consumer perceptions on seafood safety.

The loss of public trust was also exacerbated by a slow government response to reports of seafood contamination. For example, Louisiana fielded a number of claims of contaminated seafood in the wake of the spill.[79] Initially, BP provided a hotline for consumers to call and report concerns of contaminated seafood. These calls went to a call center, and consumers were assured that someone would come to collect the samples for testing. However, the BP response to these calls was sporadic, so there were many occasions where no one came to collect the samples and the seafood spoiled before tests could be run.

In interviews with community members, Commission staff was told that even when samples were collected, individuals often were never apprised of the results. Rather, the whole testing process became just another black box, leaving citizens puzzled and feeling disenfranchised. To stem consumer frustration, Louisiana eventually provided its own 24-hour hotline through the Department of Wildlife and Fisheries. These calls were always answered, according to a state official close to the process, and state officials always traveled to collect the samples for testing. However, the state hotline did not go live until September, by which time many residents were fed up.* Officials

* In describing the process instituted by Louisiana in September, Olivia Watkins of the Louisiana Department of Wildlife and Fisheries says that a state official would go to the caller's location to collect the sample, but that it was not uncommon for the caller to rescind their complaint once officials arrived on the scene. "To date, there has not been a single case of contaminated seafood brought up by any individual that was legitimate. Everything has PAHs in it – it's the level that determines whether or not [a sample] is contaminated." Watkins also says that a number of reports of contamination that initially seemed credible ended up being unrelated to the oil spill: For example, some callers reported shrimp with a black substance in their lungs but further testing revealed these shrimp were infected by

admit the delay in creating a reliable mechanism to follow-up on citizen reports of contaminated seafood contributed to the perception problem.

Beyond their own missteps in the public relations arena, authorities also blame the press for feeding the perception problem: For example, Watkins says that while reporters were quick to jump on stories of potential contamination, the same reporters often neglected to follow-up after testing was complete and the seafood in question was proved to be uncontaminated after all.

C. Injury to the Gulf Economy

Florida State University oceanographer Ian McDonald has called the Gulf of Mexico "the hardest working of our ocean basins."[80] The southern coast of the United States produces more than one-third of the nation's domestic seafood supply,[81] including most of the shrimp, crawfish, blue crabs, and oysters.[82] It produces one-third of all domestic oil,[83] claims four of the top seven trading ports by tonnage,[84] and receives runoff from nearly 40 percent of the nation's land.[85] The northern Gulf also provides diverse fish nursery and feeding grounds in the form of expansive marshes, mangrove stands, swamp forests, and seagrass beds, and boasts some of best beaches and waters in the United States for recreation and tourism.[86] Coastal tourism and commercial fisheries generate more than $40 billion of economic activity annually in the five Gulf States.[87]

In 2008, according to NOAA, Gulf commercial fishermen harvested 1.27 billion pounds of finfish and shellfish that earned $659 million in total landings revenue.[88] Other contributors to the total Gulf fishing economy are seafood processors, warehouses, distributors, and wholesalers. Further, the health of the fishing industry is intimately bound up with the wellbeing of innumerable other Gulf businesses, including dock owners, restaurants, hotels and tourism. A Louisiana survey reports that seafood-laden Louisiana cuisine is the most-cited reason for visiting the State.[89] But the health of the Gulf fishing industry is also vitally important to the nation: Gulf fishermen land 73

"black gill disease," which shrimp can catch as a matter of course. Another issue arose when consumer groups began conducting independent testing and claiming contamination. One such initiative caused a stir after local activist Nancy Mackenzie claimed to have found above-threshold PAH contamination in Louisiana shrimp: But experts say those results are not credible because the lab tested only the digestive tracts of the shrimp, which contain significantly higher PAH concentrations than shrimp muscle yet comprise a relatively insignificant fraction of the shrimp humans consume.

percent of the Nation's shrimp— half from Louisiana waters. Louisiana accounts for 67 percent of the nation's oyster production and 26 percent of the blue crab production.[90]

In testimony before the Commission in late September, Timothy Fitzgerald, a specialist on seafood safety at Environmental Defense Fund, summarized some of the economic challenges ahead:

> Based on federal and state testing, the public has been repeatedly reassured that Gulf seafood is safe to eat. However sales remain depressed in the region and around the country, and some fishermen are hesitant to return to work given poor markets and concerns over lingering toxins in the water. . . For Gulf fishermen and other members of the seafood industry, this lack of consumer confidence can be devastating. They not only face the pain of immediate financial losses, but must also struggle with the uncertainty of trying to regain the business they once had.[91]

Fitzgerald's concerns remain very much relevant more than four months later, as media reports continue to emphasize uncertain demand for Gulf seafood and related losses in tourism and market share.[92] Given the myriad factors driving consumer behavior, the true extent of the economic losses stemming from the spill-induced fisheries closures and related declines in consumer confidence may never be known. Nevertheless, a comparison of seafood and tourism revenues from 2010 to annual revenues from more prosperous times, past and future, should provide a rough indication.

It is currently not clear to what extent the enormous indirect economic impacts associated with loss of consumer confidence and injuries to the Gulf coast "brand" will ultimately be deemed compensable and that resulting uncertainty has generated intense debate among diverse government entities, local communities, interest groups, and BP. The federal Oil Pollution Act, for instance, does expressly recognize the appropriateness of compensation for "loss of profits or impairment of earning capacity resulting from property loss or natural resource injury." [93] But there is no easy legal answer to the question of how closely linked those lost profits or earnings must be to the spill before they should be deemed compensable. The search for such a rational endpoint for liability has already stymied the third Gulf Coast Claims Facility in its processing of claims.[94] The absence of clear and fair procedures for systematically evaluating such claims deserves focused attention as the lessons from the *Deepwater Horizon* oil spill are learned.

D. Impacts on Recreational and Subsistence Fishing

The *Deepwater Horizon* oil spill has also caused significant disruption to the recreational and subsistence fishing communities in the Gulf. Saltwater recreational fishing is an economically important activity in many coastal communities, contributing $41 billion dollars in economic output in the Gulf Coast region annually and supporting over 300,000 jobs.[95] It too was virtually shut down in the wake of the oil spill and is suffering a lingering brand apprehension effect, which in turn has consequences for the Gulf leisure tourism industry.[96] The seafood testing and reopening protocols have further caused confusion among subsistence fishing communities in the Gulf—many of whom do not speak English as a first language—who may not have understood or relied on public information on closure boundaries and seafood testing results. "If you can realize how difficult all this has been for the fishermen of the Gulf Coast since the day of this explosion, and then try to imagine going through it without understanding English, you will have some sense of what it has been like for the non-English proficient communities."[97]

III. Restoring Consumer Confidence: Current Efforts

The universe of Gulf seafood consumers can be conceptualized as spanning a bell-shaped curve. On the far left, you have those who continue to dispute the safety of Gulf seafood in the face of contrary evidence. These are the individuals who have dug in their heels, and most agree it makes little sense to expend additional resources trying to persuade them. On the far right of the curve, by contrast, are the individuals who trust the seafood is safe and have no compunction about enjoying the bounty of the Gulf. These are the converted. But the majority of seafood consumers are clustered closer to the middle of the bell, the midpoint between both extremes. These consumers don't have a firm opinion about whether Gulf seafood is safe, and their uncertainty offers the greatest opportunity for state and federal marketing efforts aimed at restoring confidence in the Gulf brand.

Most commercial Gulf seafood species seem to have emerged from the oil spill without any clear evidence of taint or contamination.[98] The real impact here is the reputational damage to Gulf seafood as a safe brand. Continued government testing, improvements in public outreach, and a coordinated marketing campaign are all needed to expedite its recovery. But it is also becoming increasingly apparent that time takes time—irrespective of marketing campaigns and pleas from state officials to visit their state, the consumer confidence problem has proven to be an intractable one.[99] And restaurateurs are following their customers' lead. When Commission staff asked a waiter in a prominent Washington DC seafood restaurant if their seafood was from the Gulf, the response: "We are waiting for the dust to settle."[100] There are some promising anecdotes from seafood restaurants and retailers that consumers seem to be moving beyond their apprehension,[101] but national polls conducted in late summer suggest that even the capping of the well was not sufficient to budge consumer fears, which were relatively constant since the spill began.

A. State Marketing Efforts

Even with the dawn of a new year, press reports and interviews with state officials and Gulf residents indicate that the Gulf seafood industry is far from economically restored. The most seriously affected states—Louisiana, Mississippi and Florida—have launched vigorous seafood testing and marketing campaigns aimed at rejuvenating the Gulf seafood brand. Two of these states—have requested and have received money from BP for this purpose. Alabama has requested funds from BP, which the company is considering as of January 2010. State officials in Mississippi suggest that they also have negotiations in play with BP. After several requests over several months, BP relented in early November and agreed to give Louisiana $48 million and Florida $20 million for seafood testing and marketing.[102] BP is considering a similar request from Alabama.[103] Each state's program for restoring consumer confidence is summarized below.

1. Louisiana

On May 29, 2010, Louisiana state officials sent a letter to BP, asking it to pay for a $457 million makeover of the "Louisiana brand" for injuries to consumer confidence and tourism occasioned by the oil spill. The letter highlighted the seafood industry's role as "one of Louisiana's staple economic engines," and argued that a long-term strategy grounded in marketing and better science was needed to restore eroding public confidence. In this spirit, the letter attached a proposal for a 20-year multi-agency initiative that would seek to accomplish three goals:

1. Implementation of a science-based seafood safety testing program with transparent metrics of safety and quality.
2. Implementation of a certification program for quality and processing of certified Louisiana seafood.
3. A successful short-term and sustained long-term consumer information campaign designed to reassert the Louisiana brand.[104]

After months of negotiation with BP, the State's proposal was reduced to a five-year, $173 million plan.[105] That proposal was subsequently pared down even more and, in the end, BP agreed to provide $48 million to the state over three years for seafood testing and marketing.[106] Of that amount, $18 million—or $6 million per year—will be used for seafood testing for oil, dispersant and other spill- related impacts. Independent testing would seem to be a crucial component of any plan to rebuild public confidence after the spill, as critics of the process have expressed concern that federal standards were too lax or that sampling and testing lacked transparency. Another $30 million—$10 million per year—is being provided for marketing purposes. The plan provides that the funding commitment for both programs—seafood testing and marketing—would be reset upon any oiling that would trigger the closure of additional fishing areas.

By agreement with the state, BP will provide the marketing monies to the Louisiana Fisheries Foundation—an entity associated with state government but not part of it—to ensure that the funds are not spent for purposes other than seafood marketing. The Foundation will turn the monies over to the Louisiana Seafood Promotion and Marketing Board, a quasi-independent entity which has been working on state seafood marketing campaigns since the 1990s. The Board will use the money to fund marketing efforts through state purchasing and grants. According to Watkins, the goal is to hire a firm that specializes in intense marketing and build out a strategic plan for the next three years. Watkins also mentioned the burgeoning efforts of the "National Seafood Marketing Coalition" which would go beyond the Gulf to market American seafood.*

Louisiana has tested thousands of seafood samples since the *Deepwater Horizon* spill began and will continue to do so for the foreseeable future.

* In the spirit of the times, Louisiana's public relations program also employs the use of social networks: "When [reports of contamination] came out, I was on my Tweet deck, responding to each concern as it arose." Similar trends are apparent also: Commission staff observed that in contrast to traditional op-ed pieces, state officials are using the online comment forums available for articles posted in online newspapers to address citizen concerns.

Watkins reports that the state aims to test roughly 400 samples a month going forward. At the close of 2010, 98.5 percent of Louisiana state fisheries were open. The last remaining closure was in Barataria Bay, one of the most heavily oiled areas from the spill. State officials could not say when those waters would be reopened. Going forward, Watkins says that the "number one priority is giving folks easy access to test results—providing detailed information how tests are done and letting them see the numbers showing that everything is OK."[107]

2. Mississippi

The Mississippi Departments of Marine Resources and of Environmental Quality launched a seafood safety program in September 2010 in an effort to kick-start the state's lagging seafood sector.[108] In a brochure highlighting features of the plan, officials described the development of the inter-agency seafood safety testing protocol and sampling procedures, and also published results from seafood testing carried out to date. The reported test results showed the actual levels of detected PAHs in a given sample—if detectable at all—orders of magnitude below published levels of concern. The pamphlet further highlighted the state's efforts at rebuilding public trust, which at that point included a "safe seafood assurance" workshop, dockside visits and additional monitoring and inspections of seafood facilities and operations. Yet, despite these proactive efforts and favorable test results, Jewell says that public opinion is slow to shift.

Governor Barbour had previously convened a commission of scientists and business leaders to brainstorm solutions to the persistent crisis of confidence in August.[109] The commission published its draft plan on January 14, 2011, including a number of recommendations concerning seafood safety. The report maintained that "credible seafood sampling and testing protocols have been implemented," but that "these protocols must be better communicated to the public to restore consumer confidence."[110] The report further proposed the following action items: (1) Continue long-term testing and sampling of the Gulf waters, sand beaches, and seafood to produce reliable, ongoing scientific data regarding water quality and seafood safety; (2) Continue to focus on public outreach aimed at communicating all scientific findings; (3) Hire Mississippi fishermen to assist with the long-term testing and sampling process; and (4) Begin implementation of a seafood safety marketing program, utilizing all scientific data to date regarding seafood safety.[111]

Additionally, press reports suggest that Mississippi is currently in negotiations with BP to help spur economic recovery in the state.[112] The state hopes to forge an agreement on similar terms to those already established for Louisiana and Florida. “While BP PLC has made multimillion-dollar agreements with Florida and Louisiana to help spur economic recovery, no such agreement yet exists with Mississippi,” said the Governor’s press secretary Dan Turner. But negotiations are apparently moving forward: "We expect to have an agreement with BP in the near future on assistance to promote the Gulf economy and mitigate against possible future economic losses," said Turner.”[113]

3. Florida

BP has pledged to provide Florida with $20 million dollars to assist the state in its seafood safety and marketing campaign. The money is to be evenly split between seafood testing and marketing, with $10 million committed to each effort. In a call with Commission staff in January, Paul Balthrop at the Florida Department of Agriculture and Consumer Services, Bureau of Seafood and Aquaculture Marketing, said that the state is busy crafting a marketing plan but not ready to divulge any details yet. “To be honest, we are in the very beginning of the process. It takes time to figure out how to spend $10 million responsibly—not to mention BP will be scrutinizing us. This isn’t going to be like “Brewster’s Millions,” where Richard Pryor used priceless stamps to mail letters.* We are going to spend this money responsibly.”[114]

Balthrop says that loss of consumer confidence in Florida seafood remains a significant drag on the State: “There is still a negative perception, which, in reality, is a misconception: We only had 23 miles of coastline closed—Florida was virtually open for business the entire time.” Balthrop attended the seafood safety conference in Panama City last September, which brought together representatives from each of the five Gulf States, as well as from NOAA, FDA and other key federal agencies. “One of the guys—I think he was from NOAA—said that out of 28,000 samples of finfish tested, only one came back positive [for PAH concentrations above the levels of concern].”

As regards the seafood testing component of Florida’s plan, state officials say the effort has been in high-gear since August. The state has not actually received any monies from BP yet, but is charging amounts spent on testing to a billing code per a Memorandum of Understanding with the company.

* “Brewster’s Millions” is a 1985 comedy film starring Richard Pryor, in which the protagonist has to waste $30 million in 30 days in order to inherit $300 million.

Sampling and testing was initially limited to the western Panhandle but, since November, Florida has been testing seafood collected off both coasts—the Gulf and also the Atlantic.[115] Current efforts include testing in thirteen districts, averaging about 20 samples per week. State officials collect these samples from primary processors, which can provide trip tickets from the fishermen they buy directly from, allowing for verification of sampling locations and chain of custody. Roughly 400 samples have been tested since August, only ten of which have had detectable levels of PAHs.[†] Levels for each of these ten samples have been orders of magnitude below accepted risk thresholds. The state has hosted several town hall meetings in an attempt to apprise citizens of testing procedures and frequency, as well as to respond to citizen concerns.

4. Alabama

Before he left office at the end of 2010, former Governor Bob Riley led efforts to rebuild the state's lagging seafood brand. The Governor established the Alabama Coastal Recovery Commission in the wake of the spill to propose steps to hasten the state's economic recovery. Among the commission's recommendations—published in a report in December 2010—was the creation of the Alabama Fisheries Marketing Board.[116] The Board would assume responsibility for marketing Alabama's seafood and would also fund scientific testing for contaminants resulting from the BP oil spill. The Board was not enacted before Riley left office and its fate is uncertain. Nevertheless, state officials say that they are working with the Gulf States Marine Fisheries Commission to develop marketing plans going forward and that a number of other initiatives are being considered.[117]

The state is currently in negotiations with BP to fund any such seafood marketing efforts going forward. In November, Alabama Senator Richard Shelby sent a letter to BP asking the company to provide as much money to Alabama for economic losses as it had to Louisiana: "The Deepwater Horizon spill has brought devastation to Alabama's Gulf Coast, and in particular has severely hurt our fisheries and tourism industries," wrote Shelby. The letter followed BP's announcement that it would pay $48 million to boost Louisiana's seafood industry and $30 million for tourism in the state. [118] More recent press reports suggest that BP is considering a significantly smaller award for Alabama, on the order of $9 million.[119]

† Florida is currently testing for twelve different PAHs commonly found in BP crude oil, as well as dispersant compounds. State protocols do not currently include testing for heavy metals, but state officials suggest such a test may be coming soon.

B. Nongovernmental Marketing Efforts

1. American Chefs Rally for Gulf Seafood

In September 2010, more than a dozen chefs from across the country—including the Executive Chef of the White House—toured Louisiana's seafood industry in an effort to boost confidence in Gulf seafood.[120] The trip, which was organized by White House officials and the Louisiana Seafood Marketing and Promotion Board, included visits to a shrimp boat, a processing facility and a lunch briefing about Gulf seafood safety with the head of the U.S. Food and Drug Administration. Chefs followed up with "America's Night out for Seafood" on December 1, when more than 270 restaurants from New York to Los Angeles served Gulf shrimp, oysters, crayfish, redfish and other Gulf staples.[121] The White House also participated, and White House Executive Chef Cris Comerford reported in her blog: "Since the BP *Deepwater Horizon* oil spill, consumption of Gulf seafood has dropped, which is why we at the White House are so happy to play our part in reminding Americans that Gulf seafood is not only safe – but delicious."[122]

Political consultant and Louisiana native, James Carville, has recently criticized attacks on the validity of the Gulf seafood testing.[123] He notes that all of the tests conducted by a diverse array of public and private experts—ranging from the federal EPA and FDA to state and independent laboratories—has concluded that Gulf seafood is safe. "At some point, a mountain of evidence becomes overwhelming, and I don't know how many times we have to test something ... and every independent agency, every person that's tested has come back with the same result," says Carville. "It's easy to manufacture fear, but it's hard to manufacture test results."

2. Gulf of Mexico Reef Shareholders Plan

Beyond the role for governments in mitigating the consequences of the public loss of confidence in Gulf seafood, there is a significant role for the non-profit and private sectors as well as academia. While efforts by groups like NRDC are intent on bolstering the standards of seafood safety, sometimes their work has the unintended consequence of stoking public fears.

In testimony before the Commission, EDF's Timothy Fitzgerald noted his organization's efforts on behalf of the Gulf of Mexico Reef Fish Shareholders' Alliance, a group supporting commercial fishermen in the Gulf of Mexico. Fitzgerald reported that the Alliance has been making progress to address the overfishing and poor fishery management practices that previously brought

Gulf fisheries to the brink of collapse,* but said that the lack of public trust in Gulf seafood after the oil spill "threatens to undermine the progress to date." The Alliance is presently developing a "Safe Seafood Initiative" to combat these concerns, comprised of three main pillars:

- A third-party seafood testing program that supplements that of NOAA and FDA;
- A rigorous chain-of-custody to track fish throughout the supply chain; and
- An aggressive public relations and marketing effort to highlight these efforts.

An independent third-party testing program will be an invaluable tool in combating public suspicion of government testing protocols. Additionally, EDF supports efforts to empower fishermen to conduct testing of their own catch and suggests government might play a role in this regard by providing financial and technical assistance. Particularly if independent testing corroborates government results, Gulf seafood should receive a significant boost. Verifying chain of custody from "Hook to Cook" is also important. The Alliance has already published its plan for a "Gulf Fresh Traceability System" which allows for real-time tracking of seafood throughout all stages of the supply chain.[124]

In his testimony before the Commission, Fitzgerald further emphasized the continuing role for government in testing seafood, urging authorities to "[a]rticulate and make public the details of a longterm testing plan." NOAA is taking steps towards this goal, and has indicated that it will maintain a seafood safety monitoring program in reopened areas, "continuing the collection and testing of seafood to ensure that Gulf seafood remains safe for consumers."[125] Additionally groups like EDF and NRDC support the expansion of seafood safety protocols to include tests for other contaminants in oil—such as heavy metals like arsenic, cadmium and mercury—in addition to the current analyses for PAHs and dispersants. With respect to dispersants, Fitzgerald says the

* "At its low point, [the fishermen's] fishing season had shrunk to only 54 days, and regulations forced them to throw back dead much of the fish they caught. But by working together and with federal regulators, these fishermen led the charge in transitioning their fishery to an innovative management program called 'catch shares.' Catch shares soon allowed the fishermen to fish year round while fish populations recovered – bringing better prices to fishermen and more economic stability to fishing ports." Written testimony of Timothy Fitzgerald, Environmental Defense Fund, Hearing before the National Commission, September 28, 2010.

public should receive more detailed information regarding NOAA's new chemical test to detect them in seafood and results to date: "This issue remains high on consumers' lists of concerns, despite continued assurances from FDA that they pose no health risks."[126] These common sense proposals provide a useful model for federal and state agencies seeking to restore confidence in Gulf of Mexico seafood.

C. Seafood Testing Going Forward

As the acute phase of the Gulf oil spill is receding in the public's mind, there is still much work to do. The frenzied pace of seafood testing efforts began dying down in December as agencies began the transition to a sustained longer-term monitoring effort to ensure samples continue to come back clean. States themselves began a transition too, with an eye towards focusing and intensifying marketing efforts to try to win back consumers that remain undecided. The uncertainty is fluid, with most consumers not convinced one way or the other that Gulf seafood is safe. And uncertainty can be just as damaging as doubt: As with any unknown, it is simply easier for people to select a noncontroversial alternative when strolling the grocery aisle.

In November, Louisiana officials met with FDA, EPA, NOAA and rest of Gulf States in New Orleans to discuss closures and reopenings, as well as testing methods.[127] At the top of the agenda was the kind of testing and surveillance process which would continue past the acute phase of the incident as part of a longer-term monitoring program. Though no samples from any of the Gulf States have returned from the lab with PAH concentrations above established levels of concern, everyone recognizes that testing must continue for the foreseeable future as part of the effort to restore consumer confidence. There is also a practical reason as well, given the threat of possible future contamination from unaccounted for subsea oil. NOAA is already conducting surveillance monitoring of reopened areas; those analyses are underway and will be posted soon. It remains to be seen whether the long-term testing protocol will include tests for other substances, aside from PAHs, such as heavy metals. Also agreement must be reached on the percentage of seafood samples that states will continue to send NOAA for analysis: Among the plans under discussion, states will continue to sample and test their own seafood but may be required to submit a random sample to NOAA on a periodic basis.[128]

IV. Lessons Learned

The oil spill plainly challenged existing government resources related to seafood safety: prompting an immediate need for on-the-fly interagency coordination, new protocols, public relations, and communication. Undoubtedly, the effort has been impressive, even if flawed. The interagency coordination and development of new protocols for testing seafood caught in the Gulf of Mexico has likely resulted in the most rigorously tested seafood in the world. Nonetheless, authorities have fallen short in public relations and communications arena. Slow to get ahead of public opinion, authorities have failed to go the extra mile to tackle consumer concerns.

While the success of the seafood safety monitoring and marketing schemes proposed by various Gulf States has yet to be borne out, it is apparent that loss of public confidence is rooted and resilient— neither likely to be resolved in the near-term nor susceptible to any quick fix. Nevertheless, there may be some options for speeding the healing process for this disaster as well as circumventing a similar crisis of confidence the next time. As several other Commission staff working papers have indicated, government needs to be responsive to citizen concerns from the beginning: providing clear lines of communication, updates as a situation unfolds, and open forums to air citizen concerns. At the end of 2010, the NRDC expressed concerns regarding the Gulf seafood testing protocols and were still awaiting a response from FDA and NOAA.[129]

There will always be challenges to a government's response in an emergency, and different stakeholders will have different priority areas that authorities will have to strive to balance. Nevertheless it is evident that many critical opportunities were missed this summer—occasions where concerns were raised that could have been quickly and effectively responded to, yet stakeholders felt they were ignored.[130] Commission staff did not evaluate the validity of the sampling process, which may be entirely sound. However, the fact that it has not been provided to stakeholders just clouds the process in unnecessary murk, and adds to the bad blood between government and community members. It would behoove NOAA and the states to make the details of its sampling procedures and results available as soon as possible, and in a form that could be more readily understood by citizens.

The government response on the separate issue of the PAH risk thresholds has also fallen short of quelling citizen concerns. Citizen groups reasonably requested that Gulf seafood consumption patterns be further analyzed in determining safe levels of exposure. Authorities said they were forced to use national surveys because there was no credible survey specific to Gulf coast populations. When citizen groups offered a rational alternative, namely to use scientifically validated consumption rates adopted by the EPA and the World Health Organization for high-volume seafood consumers, authorities declined. FDA ought to either commission a survey of seafood consumption patterns along the Gulf coast, or adopt the EPA and the World Health Organization for high-volume seafood consumers. And, they should include straightforward communication as to whether the consumption rates would matter in any case, given the very low PAH concentrations that were generally found in analyzed samples.

Finally, there must be a rapid-response system for addressing reports of oiling or contamination by fishermen and other coastal residents. As Olivia Watkins of the Louisiana Department of Wildlife and Fisheries made clear, citizen frustration with the lack of response to such reports fueled the debate unnecessarily in the early months of the spill. The U.S. Environmental Protection Agency used effectively an analogous program to reassure local communities about concerns they had about air pollution in the immediate aftermath o the oil spill. In the weeks after the oil spill, coastal residents were reportedly bombarded with overwhelming odors of oil depending on the direction of the prevailing winds. Meanwhile, they were being told by authorities that air quality was just fine based on testing taking place at the time. This apparent disconnect between official reports and public perception heightened the mood of public distrust. In response, EPA established a system whereby residents could call a hotline if they had concerns about air quality and EPA would send someone out immediately to monitor air quality. Such a system is similarly needed with respect claims of seafood contamination. State and federal authorities generally improved their response in this regard as the BP oil spill wore on, making consumer hotlines available and following up on those calls.[131] Nevertheless, the immediate implementation of such a rapid-response program would be useful in stemming public concerns before they reach the kind of "tipping point" observed in the Gulf spill—a point after which winning back the public trust is exceedingly difficult. In sum, any strategy for rebuilding public trust in Gulf seafood would benefit from:

(1) Better transparency: For example, NOAA should consider publishing its seafood sampling protocols and warrant their validity to the public.
(2) Better science: For example, governments should consider facilitating citizen access to independent seafood testing programs; FDA should

consider revising risk thresholds to incorporate assumptions that better track seafood consumption patterns of coastal residents.

(3) Better public relations: For example, federal and state agencies should consider creating a framework for addressing stakeholders concerns as efficiently as possible.

CONCLUSION

Seafood business interests, local community activists, national non-profit organizations, and federal and state governments prescribe different pathways out of the current seafood controversy but their voices arise out of a common concern: the potentially catastrophic harm caused by the *Deepwater Horizon* spill to the Gulf seafood industry. Federal and state government cannot solve the issue through unilateral pronouncements of safety. The crisis of public skepticism is too deep to be so easily eliminated. Federal and state officials, along with business interests, community groups and NGOs, are going to have to take a series of significant and transparent steps to earn back the consumer's trust. So long as local residents, in particular, do not believe that government officials are listening, the actual scientific merits of the government's claims are unlikely to be sufficient or generate the support of vocal dissenters. Until the government can forge cooperation and consensus on seafood safety, the public will remain skeptical.

Government officials likely could go a long way in diffusing the tensions that have emerged by engaging more effectively with members of the local community, hearing their concerns, and addressing them head-on. A critical element of any such effort will likely be assurances of long-term and ongoing monitoring of reopened fisheries. Such monitoring can demonstrate to consumers the government's commitment to remain vigilant to ensure seafood safety as the longer term impacts of the spill are better understood.

End Notes

[1] Lee Schlesinger, Florida Fish and Wildlife Conservation Commission, interview with Commission staff, December 22, 2010.

[2] Email from John Oliver, NOAA, January 18, 2010. State officials from Louisiana, Mississippi and Alabama also confirmed this result for testing of samples collected in state waters. Olivia Watkins, Louisiana Department of Wildlife and Fisheries, interview with Commission staff, December 10, 2010; Joe Jewell, Mississippi Department of Marine

Resources, interview with Commission Staff, December 9, 2010; Call with Chris Denson, Marine Resources Division of the Alabama Department of Conservation and Natural Resources, January 24, 2011. See also Editorial, "Best science says seafood safe to eat," DailyComet.com, December 15, 2010 ("[FDA] and [NOAA] have tested nearly 2,000 seafood samples for oil and chemical dispersants used to break up the spill. Officials say 99 percent tested clean of any chemicals. One percent had trace amounts of chemicals, but they were all well below the safety threshold of 100 parts per million of finfish and 500 parts per million for shrimp, crabs and oysters.").

[3] Bob Marshall, "Safety of Gulf Seafood debated 8 months after BP oil spill," *Times Picayune*, December 19, 2010.

[4] "Government Says Gulf Seafood Safe, Now Consumers Decide," *Associated Press*, August 24, 2010.

[5] Bob Marshall, "Safety of Gulf seafood debated 8 months after BP oil spill," *Times Picayune*, December 19, 2010.

[6] U.S. Census Bureau: 2007 Economic Census, Statistics by Economic Sector. Compiled data from Gulf coast counties.

[7] Ben Raines, "Researcher: Fish numbers triple after oil spill fishing closures," *Mobile Press-Register*, November 7, 2010.

[8] See, e.g., Harry R. Weber, "Gulf Oil Well Is Dead But The Pain Will Remain," *Associated Press*, September 20, 2010 ("Many people are still struggling to make ends meet with some waters still closed to fishing. Shrimpers who are allowed to fish are finding it difficult to sell their catch because of the perception — largely from people outside the region — that the seafood is not safe to eat."); Brian Skoloff, "Final oil well sealing is small comfort to suffering, anxious Gulf residents," *Associated Press*, September 20, 2010; Kim Murphy, "A poor appetite for gulf seafood: In Louisiana, where the fishing industry is as much a part of the state as Mardi Gras, even locals won't eat the catch," Los Angeles Times, August 26, 2010; Mark Albright et al., "Eight weeks after oil stopped gushing, people are still wary of gulf seafood," *St. Petersburg Times*, September 18, 2010.

[9] Kari Huus, "Panel challenges Gulf seafood safety all-clear," *MSNBC.com*, December 27, 2011; Wilma Subra, interview with Commission staff; Marilee Orr, interview with Commission staff; Tracy Kuhns, interview with Commission staff.

[10] Sammy Fretwell, "Gulf spill hurts S.C. Eateries," SunNews, com, December 12, 2010; Steve Phillips, "'Gulf Safe' seafood campaign continues," *WL OX News*, January 27, 2011.

[11] Ve Van Nguyen, interview with Commission staff, November 23, 2010; Grace Scire, interview with Commission staff, October 5, 2010.

[12] This primer on hydrocarbon bioaccumulation is courtesy of the Commission's environmental science advisor, Dr. Robert Spies.

[13] 75 Fed. Reg. 24822, May 6, 2010; 75 Fed. Reg. 26679, May 12, 2010.

[14] 75 Fed. Reg. 27217, May 14, 2010.

[15] 16 U.S.C. §§ 1801-1884.

[16] 16 U.S.C. § 1855(c).

[17] Press Release, Louisiana Department of Wildlife and Fisheries, In Precautionary Move, L.D.W.F. And D.H.H. Announce Closures Due To Oil Spill, April 29, 2010.

[18] Ibid.

[19] Press Release, Mississippi Department Of Environmental Quality, Mississippi Department Of Marine Resources, "MDEQ and DMR Continue Preparations," May 3, 2010.

[20] Press Release, Mississippi Department Of Environmental Quality, Mississippi Department Of Marine Resources, and the Mississippi State Department of Health, "Mississippi Waters Are Open, Mississippi Seafood Is Safe to Consume," May 7, 2010.

[21] Press Release, Mississippi Department of Marine Resources, "Precautionary Closure of Portions of Mississippi Marine Waters To All Commercial and Recreational Fishing, Effective Immediately," June 1, 2010

[22] The Florida Constitution delegates closure authority to the Florida Fish and Wildlife Conservation Commission.

[23] Press Release, Department of Conservation and Natural Resources, "Some State Waters Closed to Fishing," June 2, 2010.

[24] Press Release, Department of Conservation and Natural Resources, Additional State Waters Closed to Fishing, June 10, 2010.

[25] Texas Parks and Wildlife Department, *Emergency Notice: BP Oil Spill in the Gulf of Mexico*, July 28, 2010.

[26] Joe Jewell, Mississippi Department of Marine Resources, interview with Commission Staff, December 9, 2010.

[27] Ibid.

[28] Press Release, NOAA, NOAA, FDA Continue Ramping Up Efforts to Ensure Safety of Gulf of Mexico Seafood, June 14, 2010.

[29] Protecting the Public from Oil-Contaminated Seafood: Fishery Area Closure and Surveillance Plan, June 14, 2010, http://www.deepwaterhorizonresponse.com/posted/2931/NOAA_FDA_Surveillance_Plan_62_14_CLEARED_658 415.658415.pdf

[30] Press Release, NOAA, FDA Continue Ramping Up Efforts to Ensure Safety of Gulf of Mexico Seafood, June 14, 2010.

[31] Phone call with Meghan Scott, FDA, January 18, 2011.

[32] Email from John Oliver, NOAA, January 18, 2010; phone call with Meghan Scott, FDA, January 18, 2011.

[33] *Protocol for Interpretation and Use of Sensory Testing and Analytical Chemistry Results for Re-Opening Oil- Impacted Areas Closed to Seafood Harvesting* ("Reopening Protocol" or "Protocol"), June 18, 2010, *http://sero.nmfs.noaa.gov/sf/deepwater_horizon/attachment1 (3).pdf.*

[34] Reopening Protocol, June 18, 2010.

[35] Reopening Protocol, June 18, 2010.

[36] Press Release, FDA, NOAA and FDA announce chemical test for dispersant in Gulf seafood: All Samples Test Within Safety Threshold, October 29, 2010.

[37] Ibid.

[38] Ibid.

[39] Ibid.

[40] Solomon interviews; written testimony of Timothy Fitzgerald, Environmental Defense Fund, Hearing before the National Commission, September 28, 2010.

[41] See, e.g., Letter from NGO's to Donald Kraemer, Acting Deputy Director of the Center for Food Safety and Applied Nutrition, Food and Drug Administration, December 8, 2010.

[42] Joe Jewell, Mississippi Department of Marine Resources, interview with Commission Staff, December 9, 2010.

[43] Ibid. See also Olivia Watkins, Louisiana Department of Wildlife and Fisheries, interview with Commission staff, December 10, 2010.

[44] See note 50, supra, citing Office of Response and Restoration, *Oil Contaminants in Seafood: Evaluating the Risk for People Who Eat Fish and Shellfish*, June 1, 2010.

[45] Email from Albert Huang, NRDC, to Gulf residents, December 3, 2010.

[46] Letter from NGO Coalition to Food & Drug Administration, October 1, 2010.

[47] Letter from NGO Coalition to Food & Drug Administration, December 8, 2010.

[48] Gulf Coast Seafood Consumption Survey, NRDC, December 2010.

[49] Jewell interview.

[50] September 2, 2010 Re-Opening, Attachment 4: Seafood Testing Data, *http://sero.nmfs.noaa.gov/sf/deepwater_horizon/Sept2_Attachment4.pdf.*

[51] Gina Solomon, Gulf Shrimp Testing: Is a Dozen Samples in 5000 Square Miles Enough to Reassure You? (September 2, 2010), *http://switchboard.nrdc.org/blogs/gsolomon/gulf_shrimp_testing_still_not.html.*

[52] Jewell interview.

[53] Email from John Oliver, NOAA, January 4, 2010.
[54] Email from John Oliver, NOAA, January 4, 2010.
[55] Jewell interview.
[56] Interviews with community members.
[57] *NOAA, Deepwater Horizon/BP Oil Spill: Size and Percent Coverage of Fishing Area Closures Due to BP Oil Spill* (2010), http://sero.nmfs.noaa.gov/ClosureSizeandPercentCoverage.htm.
[58] Press Release, NOAA, NOAA Reopens More Than 8,000 Square Miles in the Gulf of Mexico to Fishing: 99.6 percent of federal waters now open, November 15, 2010, http://sero.nmfs.noaa.gov/media/pdfs/2010/Area%206%20and%207%20Press%20ReleaseFINAL.pdf.
[59] *Deepwater Horizon/BP Oil Spill: Federal Fisheries Closure and Other Information,* http://sero.nmfs.noaa.gov/deepwaterhorizonoilspill.htm.
[60] Press Release, Louisiana Department of Wildlife and Fisheries, LDWF Reopens Commercial Crab Fishing in Mississippi River Delta, November 8, 2010.
[61] Ibid.
[62] See, e.g., Press Release, Mississippi, MDMR and MDEQ Reopen Waters North of the Barrier Islands to Commercial and Recreational Finfish and Shrimp Fishing, July 30; 2010; Press Release, Mississippi, Mississippi Marine Territorial Waters North of the Barrier Islands Reopen to All Commercial and Recreational Blue Crab Fishing, August 20, 2010.
[63] Press Release, Department of Conservation and Natural Resources, Some Alabama Waters to Reopen for Fishing, August 7, 2010
[64] Press Release, Department of Conservation and Natural Resources, All Alabama Waters Now Open for Fishing, August 16, 2010.
[65] Executive Order, Marine Resources Division, September 5, 2010, http://www.outdooralabama.com/images/file/Shrimp%20Opening%20Sept%206%202010.pdf.
[66] Press Release, NOAA, NOAA Reopens More Than 8,000 Square Miles in the Gulf of Mexico to Fishing: 99.6 percent of federal waters now open, November 15, 2010, http://sero.nmfs.noaa.gov/media/pdfs/2010/Area%206%20and%207%20Press%20Release_FINAL.pdf.
[67] Press Release, NOAA, NOAA Closes 4,200 Square Miles of Gulf Waters to Royal Red Shrimping, November 24, 2010, http://www.noaanews.noaa.gov/stories2010/20101124_closing.html.
[68] Email from John Oliver, NOAA, January 21, 2010.
[69] Press Release, NOAA, BP Oil Spill: NOAA Re-opens Federal Waters to Royal Red Shrimp Fishing, February 1, 2011, http://sero.nmfs.noaa.gov/sf/deepwater_horizon/FB11-010ReopeningRoyalRed_020211.pdf.
[70] Press Release, NOAA, BP Oil Spill: NOAA Re-opens Federal Waters to Royal Red Shrimp Fishing, February 1, 2011, http://sero.nmfs.noaa.gov/sf/deepwater_horizon/FB11-010 Reopening_RoyalRed_020211.pdf.
[71] "Government Says Gulf Seafood Safe, Now Consumers Decide," Associated Press, August 24, 2010; "Americans Still Afraid to Eat Seafood, Poll Says," *CBS News*, August 16, 2010. See also Shaila Dewan, "Questions Linger as Shrimp Season Opens in Gulf," *New York Times*, August 16, 2010.
[72] Louisiana Office of Tourism, Effects on Perception/BP Oil Spill Survey Wave 2 Results (August 16, 2010).
[73] Testimony of Scott Angelle, Hearing before the National Commission, September 28, 2010.
[74] Office of Tourism, Effects on Perception/BP Oil Spill Survey Wave 1 Results (May 28, 2010).
[75] Ibid.
[76] University of Minnesota Food Industry Center, "Continuous Consumer Confidence in Food Safety/Defense Tracking" (presentation, FDA Symposium, September 16, 2010). The Continuous Consumer Food Safety/Defense Tracking program is a joint initiative between the University of Minnesota and the Department of Homeland Security's National Center

for Food Protection and Defense, which monitors consumer perceptions on a variety of national food issues.

77 Greater New Orleans, Inc. A Study of the Economic Impact of the Deepwater Horizon Oil Spill: Part III-Brand Damage. Forthcoming February 2011.

78 NOAA Sea Grant Agent, interview with Commission staff, September 13, 2010.

79 Watkins interview.

80 John Flesher, "Gold Rush On The Gulf: Researchers Clamor For Cash," *ABC News*, September 29, 2010.

81 Mary Rickard, "Gulf Coast oil spill chills seafood industry," *Reuters*, May 12, 2010; Kat Kinsman and Sarah LeTrent, "Gulf Coast Chefs, Fishermen Fight Tide of Misinformation," *CNN*, May 12, 2010.

82 NOAA, *NOAA's Oil Spill Response: Fish Stocks in the Gulf of Mexico* (May 12, 2010), http://www.response.restoration.noaa.gov/bookshelf/1886Fish-Stocks-Gulf-fact-sheetv2.pdf.

83 U.S. Energy Information Administration, *Gulf of Mexico Fact Sheet* (October 2010), http://www.eia.doe.gov/special/gulfofmexico/index.cfm.

84 EPA, *General Facts about the Gulf of Mexico*, http://www.epa.gov/gmpo/about/facts.html#resources.

85 National Park Service, *Mississippi River Facts* (2010), http://www.nps.gov/miss/riverfacts.htm.

86 NOAA, *The Gulf of Mexico at a Glance: A Tool for the Gulf of Mexico Alliance and the American Public* (June 2008), http://gulfofmexicoalliance.org/pdfs/gulf_glance_1008.pdf.

87 U.S. Census Bureau, *2007 Economic Census*.

88 NOAA, *NOAA's Oil Spill Response: Fish Stocks in the Gulf of Mexico* (May 12, 2010), http://www.response.restoration.noaa.gov/book_shelf/1886_Fish-Stocks-Gulf-fact-sheetv2.pdf.

89 Louisiana Office of Tourism, Effects on Perception/BP Oil Spill Survey Wave 2 Results (August 16, 2010).

90 Ibid.

91 Written testimony of Timothy Fitzgerald, Environmental Defense Fund, Hearing before the National Commission, September 28, 2010.

92 Kari Huus, "Panel challenges Gulf seafood safety all-clear," MSNBC.com, December 27, 2011; Steve Phillips, "'Gulf Safe' seafood campaign continues," *WLOX News*, January 27, 2011.

93 33 U.S.C. § 2702(b)(2)(E). The Oil Pollution Act recognizes six broad categories of damages, of which lost profits is just one: (1) injury to natural resources (recoverable by federal or state trustees); (2) loss of real or personal property and any resultant economic losses (recoverable by an owner of that property); (3) loss of subsistence use of natural resources (recoverable by a subsistence user); (4) loss of revenues resulting from destruction of property or natural resource injury (recoverable by a government claimant); (5) loss of profits or impairment of earning capacity resulting from property loss or natural resource injury (recoverable by any claimant); and (6) costs of providing extra public services during or after spill response (recoverable by a government claimant). 33 U.S.C. § 2702(b)(2).

94 David Segal, "Should BP's Money Go Where the Oil Didn't?," *New York Times*, October 23, 2010 (noting that Kenneth Feinberg has "hired one of the country's foremost scholars on torts . . . to write a memorandum about the validity and value of proximity claims.").

95 American Sportfishing Association, *New Economic Data Illustrates Gulf Oil Disaster's Significant Impact on Recreational Fishing-Dependent Businesses*, May 21, 2010.

96 See, e.g., Ed Anderson, "Leisure tourism in Louisiana is expected to take a hit after BP oil spill," the *Times- Picayune*, January 23, 2011.

97 Grace Scire, Boat People SOS, email correspondence with Commission staff, September 17, 2010.

[98] See, e.g., Press Release, NOAA, NOAA and FDA Announce Chemical Test for Dispersant in Gulf Seafood; All Samples Test Within Safety Threshold, October 29, 2010, *http://www.noaanews.noaa.gov/stories2010/20101029_seafood.html.*

[99] Chris Kirkham, "Convincing public that Gulf of Mexico seafood is safe will take time, experts say, *Times Picayune*, August 8, 2010.

[100] Discussion with wait staff at the Oceanaire Seafood Room, Washington, DC, January 11, 2011.

[101] "Government Says Gulf Seafood Safe, Now Consumers Decide," Associated Press, August 24, 2010.

[102] Press Release, BP, Seafood Safety, Tourism and Coastal Restoration Funding Announced, November 1, 2010, http://www.louisianagulfresponse.com/go/doc/3047/940587/; Press Release, Florida Department of Agriculture and Consumer Services, Bronson Announces That BP Will Pay $20 Million To Fund Seafood Inspections, Marketing Efforts In Wake Of Oil Spill, October 25, 2010. *http://www.doacs.state.fl.us/press/2010/10252010_2.html.*

[103] Paul Quinlan, "Gulf Spill: Ala., La. fighting over potential billions in BP penalties," *Energy & Environment Daily*, November 23, 2010.

[104] Ibid. See also Louisiana Seafood Safety Response and Quality Certification Plan: Post-Mississippi Canyon 252 Oil Spill, May 28, 2010.

[105] See Louisiana Seafood Safety Response and Certification Plan, http://www.wlf.louisiana.gov/louisiana-seafoodsafety-response-and-certification-plan.

[106] "Jindal Announces Agreement with BP for Seafood Safety, Coastal Restoration & Tourism Funding," *Baton Rouge Today*, November 1, 2010.

[107] Watkins interview.

[108] Mississippi Department of Marine Resources, *Mississippi Seafood Safety*, September 2010.

[109] Press Release, Mississippi, Governor Barbour Establishes Panel to Set Vision for Gulf Coast Recovery, Development, August 16, 2011.

[110] The Mississippi Gulf of Mexico Commission, *Draft Vision for Gulf Coast Recovery, Restoration, and Protection* (January 14, 2011), 15.

[111] Ibid. at 21.

[112] George Altman, "Mississippi leaders wary of federal oil spill report," Mississippi Press, January 13, 2011, http://blog.gulflive.com/mississippi-press-news/2011/01/mississippi_leaders_wary_of_fe.html.

[113] Ibid.

[114] Call with Paul Balthrop, Florida Department of Agriculture and Consumer Services, Bureau of Seafood and Aquaculture Marketing, January 14, 2011.

[115] Call with Richard Stevens, Administrator of Sanitation and Safety, Florida Dept of Agriculture, Division of Food Safety, January 15, 2011.

[116] Russ Henderson, "Riley to create seafood marketing board," *Mobile Press-Register*, December 18, 2010.

[117] Call with Chris Denson, Marine Resources Division of the Alabama Department of Conservation and Natural Resources, January 24, 2011.

[118] Paul Quinlan, "GULF SPILL: Ala., La. fighting over potential billions in BP penalties," *Environment & Energy Daily*, November 23, 2010.

[119] Russ Henderson, "Riley to create seafood marketing board," *Mobile Press-Register*, December 18, 2010 ("The state remains in negotiations with BP to obtain funds for economic recovery, Stacy said. Riley said Wednesday that the marketing effort could get $9 million from BP.").

[120] Press Release, Louisiana Seafood Promotion and Marketing Board, Top U.S. Chefs Focus on Importance of Gulf Seafood, September 13, 2010.

[121] Press Release, BP Florida Gulf Response, Restaurants, White House Promote Safety of Gulf Seafood, December 3, 2010.

[122] Cris Comerford, "A Taste of the Gulf at the White House," White House Blog, Decmeber1, 2010, http://www.whitehouse.gov/blog/2010/12/01/a-taste-gulf-white-house.

[123] Press release, Louisiana Seafood Promotion and Marketing Board, James Carville Takes Issue with Gulf Seafood Critics, January 17, 2010.

[124] Gulf of Mexico Reef Fish Shareholders' *Alliance, Safe Seafood Initiative*, http://shareholdersalliance.org/sustainability/safe-seafood-initiative/.

[125] Press release, NOAA, NOAA Strategy for Future Reopenings: NOAA seafood sampling and testing priorities for federal closed area, September 27, 2010.

[126] Written testimony of Timothy Fitzgerald, Environmental Defense Fund, Hearing before the National Commission, September 28, 2010.

[127] Watkins interview.

[128] Ibid.

[129] Email with Gina Solomon, December 29, 2010.

[130] Interviews with community leaders at the Weeks Bay Conference, October 4-6, 2010, Weeks Bay, Alabama.

[131] Watkins interview.

In: Impacts of the Gulf Oil Spill on Fishing ... ISBN: 978-1-61470-971-8
Editors: C. Scarnati and E. Popeo

Chapter 4

COMMERCIAL FISHERY DISASTER ASSISTANCE*

Harold F. Upton

SUMMARY

Disaster relief may be provided by the federal government to assist the fishing industry when it is affected by a commercial fishery failure. A commercial fishery failure occurs when fishermen endure economic hardships resulting from fish population declines or other disruptions to the fishery. The Department of Commerce can provide disaster assistance under Sections 308(b) and 308(d) of the Interjurisdictional Fisheries Act (16 U.S.C. § 4107), as amended, and Sections 312(a) and 315 of the Magnuson-Stevens Fishery Conservation and Management Act (16 U.S.C § 1861). The National Marine Fisheries Service plays a central role in determining whether a commercial fishery failure has occurred and in allocating federal funding to states and affected fishing communities. Congress plays a pivotal role by appropriating funds and providing oversight of the process.

Fisheries are subject to environmental variability that may affect the fishery resource and/or commercial infrastructure such as boats, shoreside processing, and ports. Since 1994, federal fishery failures have been declared

* This is an edited, reformatted and augmented version of Congressional Research Service Report RL34209, dated February 22, 2011.

on 33 occasions, and nearly $860 million in federal funding has been appropriated for fishery disaster relief. Funds have been allocated to fisheries of the North Pacific, Pacific Northwest, Gulf of Mexico, and the East Coast. The most recent fishery failures were declared for the Maine shellfish fishery and for continuation of the West Coast salmon troll fishery.

Direct federal financial assistance has been provided to fishermen and fishing communities in the form of grants, job retraining, employment, and low interest loans. Assistance has also included fishery data collection, resource restoration, research, and fishing capacity reduction programs to prevent or lessen the effects of future disruptions to fisheries. However, critics contend that disaster assistance programs often fall short of expectations because sometimes funds are not disbursed in a timely manner, ambiguities complicate the definition of a fishery failure, relief may not be integrated with long-term fishery management objectives, and funds may not reach the people who are in the greatest need of assistance. The National Oceanic and Atmospheric Administration has proposed regulations to clarify and interpret the fishery disaster assistance provisions of the Magnuson-Stevens Fishery Conservation and Management Act and the Interjurisdictional Fisheries Act, but the final rule has not been published.

INTRODUCTION

Disaster relief may be provided by the federal government to assist the fishing industry when it is affected by a commercial fishery failure. A commercial fishery failure occurs when fishermen endure economic hardships resulting from fish population declines or other disruptions to the fishery. The Department of Commerce can provide disaster assistance .under Sections 308(b) and 308(d) of the Interjurisdictional Fisheries Act (IFA; 16 U.S.C. § 4107), as amended, and Sections 312(a) and 315 of the Magnuson-Stevens Fishery Conservation and Management Act (MSFCMA; 16 U.S.C § 1861). The National Marine Fisheries Service (NMFS) plays a central role in determining whether a commercial fishery failure has occurred and in allocating federal funding to states and affected fishing communities. Congress plays a pivotal role by appropriating funds and providing oversight of the process.

Table 1. Fishery Failure Causes, Types of Assistance, and Use of Funds

Section	Commercial Fishery Failure Causes	Types of Assistance and Use
Section 312(a) of MSFCMA	Fishery resource disaster as a result of— (1) natural causes (2) man-made causes beyond the control of fishery managers to mitigate through conservation and management measures, including regulatory restrictions imposed to protect human health or the marine environment (3) undetermined causes	(1) assessment of the social and economic effects of the failure (2) assistance to the community and fishermen (3) projects to restore the fishery or prevent reoccurrence of a similar failure (4) federal share of assistance cannot be greater than 75%
Section 308(b) of IFA	Fishery resource disaster arising from— (1) natural causes (2) undetermined causes	(1) restore a fishery affected by a fishery failure (2) prevent a future fishery failure (3) federal share of funding is limited to 75% of costs
Section 308(d) of IFA	Fishery resource disaster arising from— (1) natural disasters such as a hurricane or other natural disaster	(1) direct assistance to fishermen; (2) indirect assistance through state agencies, local government, and nonprofit organizations (3) no limit on the federal share of costs
Section 315 of MSFCMA	Regional fishery disaster— (1) results in economic losses to the coastal or fishing communities (2) affects more than one state or a major fishery managed by a Council or interstate fishery commission (3) is determined by the Secretary to be a commercial fishery failure under § 312(a) of MSFCMA or fishery resource disaster under § 308(d) of IFA	(1) activities authorized under either MSFCMA or IFA (2) the Secretary may waive matching requirements if no reasonable means are available for meeting the match and the probable benefit of federal financing outweighs the public interest in imposing the match

Fisheries are subject to environmental variability that may affect the fishery resource and/or commercial infrastructure such as boats, shoreside processing, and ports. Since 1994, federal fishery failures have been declared on 33 occasions, and nearly $860 million in federal funding has been appropriated for fishery disaster relief. Funds have been allocated to fisheries of the North Pacific, Pacific Northwest, Gulf of Mexico, and the East Coast.

Direct financial assistance has been provided to fishermen and fishing communities in the form of grants, job retraining, employment, and low interest loans. Assistance has also included fishery data collection, resource restoration, research, and fishing capacity reduction programs to prevent or lessen the effects of future disruptions to fisheries. Several issues related to fishery disaster relief include timing relief disbursements to meet critical needs, integrating relief with long-term management objectives, defining and declaring a fishery failure, and reaching people who may be in the greatest need of relief. The National Oceanic and Atmospheric Administration has proposed regulations to clarify and interpret the fishery disaster assistance provisions of MSFCMA and IFA. The proposal has not been finalized.

PROGRAM REQUIREMENTS AND PROCEDURES

The Department of Commerce can provide disaster assistance under either the MSFCMA or the IFA.[1] Differences exist under each law with regard to the causes of a fishery failure, and the use of funds (see **Table 1**). Several recent fishery failures have been declared under both laws, providing program managers greater latitude in matching relief with the needs of recipients.

MSFCMA

In 1996, MSFCMA was amended to include a new section focusing on transition to sustainable fisheries. This section includes Subsection 312(a) to provide fishery disaster relief when commercial fishery failures occur, especially for those fisheries in need of stock rebuilding. Under Section 312(a), the process is started at the discretion of the Secretary of Commerce, at the request of the governor of an affected state, or at the request of a fishing community representative. The Secretary determines whether a commercial fishery failure has occurred depending on three factors. First, there must be a fishery resource disaster resulting from a decrease in fish population biomass or the loss of fishing vessels, gear, or related infrastructure. Second, the cause of the fishery resource disaster must be one of the following:

- natural causes;
- man-made causes beyond the control of fishery managers to mitigate through conservation and management measures, including regulatory restrictions imposed to protect human health or the marine environment; or
- undetermined causes.

Finally, there much be an economic impact resulting from the commercial fishery disaster. Requests usually contain information describing the alleged fishery failure. Although guidelines for handling requests are not codified in rule, the Secretary typically directs the appropriate Regional Administrator for NMFS to collect and analyze required information such as the historical context, the biological and economic magnitude of the disaster, and the relationship between underlying causes and the alleged fishery disaster.[2] Depending on the circumstances, the analysis should be conducted in consultation with state(s) and should consider supporting information and data that the state(s) provide. The Secretary uses the information to determine whether or not a fishery resource disaster has occurred. The cause of the disaster must also be determined to assess eligibility under different sections of the MSFCMA and the IFA. Once it is concluded that a fishery resource disaster has occurred, fishery socioeconomic data are reviewed to determine whether a commercial fishery failure exists. The decision depends on whether a significant number of people engaged in the fishery have suffered severe economic hardship as a result of the fishery resource disaster. Once it is determined that a commercial fishery failure exists, Congress may use the authorization in the MSFCMA to appropriate funds for financial assistance to harvesters and other affected parties.

After funds are appropriated, the affected state, community, or group develops a spending plan that is evaluated by NMFS regional offices. Funding under the MSFCMA may be used to address a broad variety of needs including assessment of the social and economic effects of the failure, assistance to the community, and projects to restore the fishery or prevent reoccurrence of a similar failure. Before releasing funds, the Secretary must also determine that activities would not expand the size and scope of the failure in that fishery, other fisheries, or affect fisheries in other geographic regions. The federal share of assistance carried out under Section 312(a) of the MSFCMA cannot be greater than 75% of the cost of relief activities.

IFA

The IFA was enacted in 1986 to distribute federal funds to states for developing interstate fishery research programs. Under IFA, funds are authorized to provide assistance for a commercial fishery failure in Section 308(b) or harm caused according to Section 308(d). Under Section 308(b), the causes of a commercial fishery failure or serious disruption to future production due to a fishery resource disaster include natural and undetermined causes. In Section 308(d), fishery resource disasters are referred to as natural disasters. The definition of a fishery resource disaster appears to be broader under the MSFCMA because human-related causes are also included. Otherwise, the process of collecting information and determining whether a fishery resource disaster and corresponding commercial fishery failure have occurred is similar under Section 3 08(b) of the IFA. Instead of assessing the occurrence of a commercial fishery failure, Section 3 08(d) of the IFA requires demonstration of harm. Harm is defined as uninsured damage to fishing vessels, fishing gear, processing facilities, marketability, habitat, or infrastructure.

IFA funding under Section 308(b) may be used by states alone or by the Secretary in cooperation with the states. Funding may be provided for any purpose the Secretary determines as appropriate to restore a fishery affected by a commercial fishery failure or to prevent a future fishery failure. Under Section 308(b), funds may not be used for grants to charter fishing vessels, and the federal share of activity funding is limited to 75% of costs. Funding under Section 308(d) of IFA may be used to provide direct assistance to fishermen or to provide assistance indirectly through state agencies, local government, and nonprofit organizations. In contrast to the MSFCMA and Section 308(b) of IFA, there is no limit on the federal share of costs under Section 308(d). Section 308(d) also outlines the conditions under which funding may be used for other activities such as fishing capacity reduction programs. These programs include fishing vessel buybacks, gear reduction, or fishing permit retirement. Funding under both MSFCMA and IFA is usually appropriated by Congress as needs arise, rather than in anticipation of future needs.

MSFCMA Regional Coastal Disaster Assistance

In 2006, MSFCMA was amended by adding Section 315 the Regional Coastal Disaster Assistance, Transition, and Recovery Program. When a catastrophic regional fishery disaster occurs, the Secretary may establish a regional program to provide immediate disaster relief assistance to fishermen, charter fishing operators, U.S. fish processors, and owners of related fishery infrastructure. A catastrophic regional fishery disaster is defined as a natural disaster, such as a hurricane or tsunami, or a regulatory closure to protect human health or the marine environment. A catastrophic regional fishery disaster is an event that:

- results in economic losses to the coastal or fishing communities;
- affects more than one state or a major fishery managed by a Council[3] or interstate fishery commission; and
- is determined by the Secretary to be a commercial fishery failure under Section 312(a) of MSFCMA or as a fishery resource disaster under Section 3 08(d) of IFA of 1986.

Within two months after a catastrophic regional fishery disaster, the Secretary is required to provide the governor of each participating state with a comprehensive economic and socioeconomic evaluation of the region's fisheries. The evaluation would assess the current and future economic viability of affected fisheries including the economic impact of foreign fish imports and direct, indirect, or environmental impacts of the disaster on the fishery and coastal communities. Subject to the availability of appropriations, the program would provide funds for infrastructure needs, job training assistance, fishing capacity reduction, and for other activities authorized under either MSFCMA or IFA. Under the Regional Coastal Disaster Assistance, Transition, and Recovery Program, the Secretary may waive the matching requirements if no reasonable means are available for meeting the match, and the probable benefit of 100% federal financing outweighs the public interest in imposing the match.

Other Potential Sources of Assistance

When businesses suffer economic injuries from a disaster, the Small Business Administration (SBA) may also determine whether a disaster declaration is warranted.[4] For example, when red tide required closure of the Maine shellfish fishery in 2005, SBA evaluated the impact on small businesses and determined a disaster declaration was justified. The declaration makes affected businesses eligible for Economic Injury Disaster Loans.[5] The purpose of the loan program is to provide working capital at low interest rates to assist recovery of businesses harmed by a disaster.

The Economic Development Administration (EDA) provides community grants and revolving loan funds to help distressed communities.[6] EDA has assisted fishing communities through its Public Works Program by funding port and harbor improvements. EDA's Economic Adjustment Program helps communities adjust to economic disruptions through support of business development, planning, and market research. Industries that have been adversely affected by increased imports of similar or competitive goods can seek technical assistance under EDA's Trade Adjustment Assistance Program.

FISHERY DISASTER DECLARATIONS

Since 1994, the Secretary of Commerce has declared 33 fishery resource disasters. During this period Congress has appropriated nearly $860 million for fishery disaster relief. **Table 2** provides a list of fishery failures, the date the disaster was determined, and funds appropriated by Congress for each.

Fishery resource disasters are diverse with respect to their causes and scope. Most declarations have resulted from natural events such as hurricanes, floods, changes in ocean conditions, or algal blooms such as red tide. In coastal areas hurricanes may damage fishing industry infrastructure such as vessels, docks, fish houses, and related businesses. Even if the resource remains abundant, harvesting, processing, and transport to markets may not be possible until repairs are undertaken and basic services are restored. In addition to the costs of repairs and the replacement of equipment and gear, lost fishing time also can be costly. The fishery resource also may be directly affected if, in addition to damaged infrastructure, hurricanes cause damage to oyster beds from silt and debris. Algal blooms such as red tide are another type of natural event that can render seafood toxic and result in fishery closures. Under these conditions, fishermen may be completely shut down for months until toxin levels in shellfish decline to acceptable levels.

Declines in fishery resource abundance may result from several factors, such as natural environmental variations, human effects on the environment, and overfishing. Salmon fisheries are sensitive to natural changes in oceanic conditions. Salmon abundance has also been affected where dams, irrigation, grazing, mining, and forestry practices have degraded salmon habitat, especially for salmon populations in the Pacific Northwest. Overfishing by itself is not an acceptable cause of a fishery resource disaster because it is not beyond the control of fishery managers.[7] However, a fishery resource disaster of natural or undetermined causes may be exacerbated by overfishing. In these cases assistance may include efforts to rationalize (decrease) fishing capacity. Overfishing has also contributed to fish population declines in several resource disaster cases such as the New England multispecies fishery and the Pacific groundfish fishery. In these cases, fish abundance decreased significantly and stock rebuilding has required substantial decreases in harvest.

Table 2. Fishery Failure Declarations Since 1994

Fishery or Region	Authority	Declared	Appropriation[a]
New England Multispecies I	IFA 308(b)	3/18/94	$86.8 million[b]
Northwest Salmon Fisheries I	IFA 308(d)	5/26/94	$12 million
Gulf of Mexico Hurricanes	IFA 308(d)	8/2/95	$15 million
New England Multispecies II	IFA 308(d)	8/12/95	$26 million
Northwest Salmon Fisheries II	IFA 308(d)	8/2/95	$13 million
Bristol Bay/Kuskokwim River (AK)	MSA 312(a)	11/5/97	$7 million
Gulf of Mexico Flooding Events	MSA 312(a)	8/7/98	$3.5 million
Northwest Salmon Fisheries III	MSA 312(a)	8/7/98	$3.5 million
Bristol Bay/Kuskokwim River (AK)	MSA 312(a)	9/9/98	$50 million
Florida Trap Fisheries	MSA 312(a)	9/20/99	$4.8 million
North Carolina Fisheries	MSA 312(a)	9/22/99	$6 million
Long Island Sound Lobster	MSA 312(a)	2/4/00	$13.9 million
West Coast Groundfish Fisheries	MSA 312(a)	2/4/00	$5 million
Bering Sea Alaska Snow Crab	MSA 312(a)	5/11/00	$10 million
Alaska Salmon (Norton Sound)	IFA 308(b) MSA 312(a)	8/4/00	$15 million $7.5 million
Fraser River/Lummi Indian Salmon	MSA 312(a)	11/13/02	
Georgia Blue Crab	MSA 312(a)	5/8/03	None to date
Red Tide (Massachusetts)	MSA 312(a)	6/16/05	$2 million
Red Tide (Maine)	MSA 312(a)	6/23/05	$2 million

Table 2. (Continued)

Fishery or Region	Authority	Declared	Appropriation[a]
Gulf of Mexico Fisheries (Katrina and Rita)	MSA 312(a) IFA 308(d)	9/9/05 10/4/05	$128 million $110 million
Klamath river Basin (Salmon)	IFA 308(b) MSA 312(a)	8/10/06	$60.4 million
Sacramento River (Chinook troll salmon)	IFA 308(b) MSA 312(a)	5/1/08	$170 million
Gulf of Mexico (Gustav and Ike)	IFA 308(d)	9/17/08	$47 million
Chesapeake Bay Blue Crab (MD and VA)	MSA 312(a)	9/23/08	$20 million
Red Tide (Maine, NH, and Mass.)	MSA 312(a)	11/14/08	$5 million
Puget Sound Sockeye Salmon	MSA 312(a)	11/14/08	$2 million
Sacramento River (Chinook troll salmon)	IFA 308(d) MSA 312(a)	4/30/09	Continued from 2008
Yukon River, AK (Chinook salmon)	MSA 312(a)	1/15/10	$5 million
Gulf of Mexico (Deepwater Horizon oil spill) (FL, AL, MS, and LA)	MSA 312(a)	5/24/10	$28 million
Sacramento River (Chinook troll salmon)	IFA 308(d) MSA 312(a)	9/2/10	None to date
Red Tide (Shellfish in Maine)	MSA 312(a)	12/22/10	None to date

Source: Adapted from the NOAA, Office of Management and Budget, Fishery Disaster Assistance Web page, http://www.nmfs.noaa.gov/mb/financial_services/disaster.htm.

a. Appropriations often follow fishery failure determinations by months, sometimes in the following year.

b. Funding was appropriated on several different occasions from 1994 to 1999.

Funds for disaster assistance have been used for a wide variety of purposes, and may include direct assistance to fishermen such as:

- compensation;
- community grants;
- training;
- loans and debt refinancing; and
- employment on fishery related projects.

Table 3. Assistance Provided for Commercial Fishery Resource Disasters by Fishery Disaster and Year of Appropriation

New England Multispecies **1994—$30 million.** Assistance: fishing industry grants that included employment for fishermen (training, new business opportunities, aquaculture, marketing, and by-catch reduction), demonstration buyback program, loan program, and family assistance centers. **1995—$25 million.** Assistance: vessel buyback, administration, and fisherman health program. **1999—$6.8 million.** Continuation from 1994 failure with assistance that included compensation for lost fishing time and cooperative research. **2000—$25 million.** Continuation from 1994 failure with assistance that included permit buyback and cooperative research. **2001—$1 million.** Continuation from 1995 of the fisherman health program. **2008—$ 13.4 million.** (Disaster not declared.) Assistance: funding for fishermen, fishing businesses, and a health insurance program.
Pacific Northwest Salmon **1994—$ 12 million.** Assistance: fishing permit buyback, habitat restoration jobs, and data collection jobs. **1995—$ 13 million.** Assistance: fishing permit buyback, habitat restoration jobs, and data collection jobs. **1998—$3.5 million.** Assistance: fishing buyback program. **2007—$60.4 million.** (Klamath River-related.) Assistance: direct payments to fishermen for business expenses. **2008—$ 170 million.** Assistance: direct payments to commercial and recreational charter fishermen. **2009—$53.1 million.** (Remainder of $170 appropriated by Congress in 2008.) Assistance: direct payments to commercial and recreational charter fishermen, and related businesses. Determination of a commercial fishery failure was continued in 2010, but no funding has been provided.
Gulf of Mexico Hurricanes **1995—$15 million.** (Hurricanes and tropical storms 1992-1995.) Assistance: compensation to fishermen, Gulf states for research and habitat restoration (inshore license buyback TX and cooperative research LA). **2006—$ 128 million.** Assistance: rehabilitating oyster beds and shrimp grounds, reseeding, rehabilitating, and storing oyster reefs, and cooperative research and monitoring. **2007—$ 110 million.** Assistance similar to 2006 funding. **2008—$47 million.** Assistance: restore damaged oyster reefs, removal of storm debris, rebuilding of processing houses, docks, ice houses, and other parts of fishery related infrastructure.

Table 3. (Continued)

Alaska Salmon **1998—$7 million** (Bristol Bay/Kuskokwim River.) Assistance: community grants, loan program, economic planning grants, and fisheries research, education, and training grants. **1999—$50 million** (Bristol Bay/Kuskokwim River/Yukon River.) Assistance: emergency assistance to affected families, direct loans, community development activities. **2000—$ 15 million** (Norton Sound/Kuskokwim/Yukon River.) Assistance: economic development and loans. **200 1—$7.5 million** (Norton Sound/Kuskokwim/Yukon River.) Assistance: economic development and loans.
Gulf of Mexico Flooding **1997—$3.5 million.** Assistance: research and data collection.
Florida Trap Fishery **2000—$ 4.8 million.** Assistance: direct assistance to fishermen, buyback trap certificates, retrieve lost traps and debris, and research ongoing trap reduction program.
North Carolina Fisheries **2000—$6 million.** Assistance: direct economic relief to seafood dealers, charter and head boats and commercial fishing piers, research and resource assessment, and mitigation of oyster losses by enhancing habitat.
Long Island Sound Lobster Fishery **2000—$ 13.9 million.** Assistance: economic compensation, trap tag buyback, job training, small business development, interest subsidy loans, and research on causes of the disaster.
West Coast Groundfish Fisheries **2000—$5 million.** Assistance: compensation to individuals, provided direct sustaining aid to fishermen, and assistance to resource dependent communities.
Bering Sea Alaska Snow Crab **2000—$10 million.** Assistance: community and economic development, Bering Sea ecosystem research, and cooperative research.
Red Tide (New England) **2006—$4 million.** Assistance: pay compensation to individuals and improve management of future outbreaks (research and monitoring). **2008—$5 million.** Assistance similar to 2006. Declaration made in 2010, but no funds have been provided.
Chesapeake Bay Blue Crab (Maryland and Virginia) **2008—$20 million.** Assistance: habitat improvement, employing watermen, industry projects and diversification, and research and monitoring.

Georgia Blue Crab Declaration made in 2003, but no funding has been provided.
Fraser River/Lummi Indian Fishery (Sockeye salmon) Declaration made in 2002, but no funds were appropriated. **2008—$2 million.** Assistance: relief for tribal and non-tribal fishermen.
Yukon River Alaska **2010—$5 million.** Assistance to fishermen and communities.
Gulf of Mexico Oil Spill **2010—$28 million.** Strategic marketing plan and a health safety assurance program for Gulf of Mexico seafood ($15 million). Remaining $13 million to be released only if resources provided under other authorities are insufficient.

Source: Adapted from the National Oceanic and Atmospheric Administration, Office of Management and Budget, Fishery Disaster Assistance Web page, *http://www.* nmfs.noaa.gov/mb/financial_services/disaster.htm.

Other forms of indirect fishery-related assistance have included fishing capacity reduction (vessel, permit, and gear buybacks), formation of a fisheries research trust, economic planning grants, and research grants. **Table 3** summarizes funding and activities by fishery or disaster event for fishing disaster declarations.

State Role

States are frequently an active partner throughout the process, from requesting the Secretary of Commerce to declare a fishery failure and providing related data to disbursing relief to fishermen and related businesses. Relief funding is often provided directly to states, or in cases of regional disasters through regional commissions such as the Pacific States Marine Fisheries Commission. For example, in 2007, distribution of Oregon salmon troll fishery relief was planned and coordinated by the state's department of agriculture in cooperation with related agencies and nonprofit organizations such as the Oregon Salmon Commission. In addition to matching funds, state government may also provide funding when federal funds are not available, although historically such funding has been limited.

Fishing Capacity Reduction Programs

Many U.S. fisheries are overcapitalized—investments in fishing capacity are greater than that needed to harvest the fishery resource on a sustainable basis. When fishery resources decline precipitously, as in the case of a fishery failure, effects on the fishing industry are likely to be greater when there is excess fishing capacity operating in the fishery. First, when excess fishing capacity exists, overfishing often occurs and management goals are likely to involve rebuilding of fish populations. During rebuilding, the fishery is likely to be highly regulated with relatively low allowable harvests. Second, since many fisheries are already overcapitalized and fully exploited, there are few alternative fishing opportunities. Finally, the financial effects of any fishery failure are likely to be greater when there is overcapacity because of the larger number and/or size of vessels and associated crew participating in the fishery.

Fishing capacity reduction, often referred to as buyback programs, has been a prominent feature of several disaster relief programs. Capacity reduction is usually accomplished through the direct purchase of fishing vessels, gear, and/or fishing permits.[8] Capacity reduction is discussed in Section 312(b) of the MSFCMA and Section 308(d) of the IFA.

The general objectives of buyback programs are to provide immediate relief to fishermen, decrease the level of fishing effort to improve the profitability of the remaining fishing fleet, and conserve the resource. The effectiveness of buyback programs in reducing fishing capacity depends on whether the remaining fishermen have the incentive to continue investing in boats and gear. Often there is also "latent" fishing effort—boats and gear with permits to fish that are inactive or only marginally utilized in the fishery. The exit of some vessels may encourage this latent fishing effort (vessels) to re-enter the fishery, resulting in little or no net reduction in fishing capacity. Furthermore, the first to accept buybacks may be the least efficient vessels in the fleet. This results in fleet reductions that are relatively modest yet expensive because only the oldest and least efficient units are taken out of production.

Although capacity reduction programs attempt to provide long-term benefits to those who decide to remain in the fishery, poorly crafted programs may result in little or no benefit at the expense of taxpayers. Although a means to ease financial hardship caused by a fishing disaster, lasting benefits may depend on better recognition of the motivations of vessel owners and fishermen.

Recent Actions by NOAA and Congress

The following summaries include fishery disaster declarations since 2006. The most recent disaster determinations include a continuation of the California salmon troll fishery failure and the Maine shellfish fishery failure. Recent congressional appropriations for fishery disaster assistance have included \$28 million for Gulf of Mexico fisheries and \$5 million for the Yukon king salmon fishery.

West Coast Salmon Ocean Troll Fishery (Klamath)

On July 6, 2006, a fishery failure was declared for the West Coast ocean troll salmon fishery. Chinook salmon stocks that spawn in California and Oregon rivers intermingle in the ocean and are harvested together off the coasts of these states. Klamath River fall Chinook salmon is a key stock with respect to both landings and regulation of the fishery.[9]

The ocean troll salmon fishing season between Cape Falcon, Oregon, and Point Sur, California, was strictly limited during the 2006 season.[10] From 2001 to 2005, drought conditions in the upper Klamath Basin resulted in very low flow conditions in the Klamath River and its tributaries. Low flows likely contributed to substantial mortality of juvenile and adult Chinook salmon by creating an environment in which they become more susceptible to endemic diseases. In 2004 and 2005, returns of Klamath River fall Chinook fell below 35,000, the regulatory floor set for any one year, and in 2006, the run size was projected to be approximately 25,000. As a result of the anticipated low spawning return, the Pacific Fishery Management Council (PFMC) recommended, and NOAA issued, a Temporary Rule for Emergency Action to strictly curtail the troll salmon fishery off Oregon and California from May 1, 2006, to August 31, 2006. Although a complete closure of the fishery was avoided, landings decreased in 2006 by 81% when compared to the average of the preceding five years.

The governors of Oregon and California requested action based on the 2006 forecast of Klamath River fall Chinook salmon returns and the actions taken in the spring of 2006 by the PFMC and NMFS. Since the PFMC developed the 2006 season regulations in the spring of 2006, the likely effects of the curtailed fishery were anticipated before the actual losses were realized. Fishermen and others associated with the fishing industry were concerned that aid to fishing communities might be delayed. On July 6, 2006, the Secretary of Commerce declared a commercial fishery failure under Section 308(b) of the IFA, and on August 10, 2006, under Section 312(a) of the MSFCMA. Fishing industry concerns increased during the fall of 2006 and spring of 2007 when no federal funding was provided. In May 2007, the U.S. Troop Readiness, Veterans' Care, Katrina Recovery, and Iraq Accountability Appropriations Act, 2007, (P.L. 110-28) allocated $60.4 million to the NOAA "Operations, Research, and Facilities" account to be distributed among eligible recipients affected by the commercial fishery failure. Assistance was distributed by the Pacific States Marine Fisheries Commission to Oregon and California fishermen and Indian tribes that rely on salmon. Oregon salmon troll fishery landings and revenue improved only slightly during the 2007 season. In 2008, the ocean fishery was limited by low Chinook salmon returns to the Sacramento River.

New England Multispecies Fishery

In 2007, the governors of Maine, Massachusetts, and Rhode Island requested that the Secretary of Commerce declare a commercial fishery failure for the Northeast Multispecies (groundfish) fishery. They cited economic hardships endured by New England fishermen because of restrictive fishery regulations for groundfish species such as cod. In October 2007, NMFS responded that revenue declines in Maine and Massachusetts were not sufficient to warrant a commercial fishery failure. NMFS cited increases in 14 of 18 groundfish stocks in the most recent stock assessment and total fishery revenue increases for some ports during the previous year. Industry representatives responded that a disaster was declared 13 years ago when fish landings were more than twice as high as in 2007. The actual biological and economic impacts cited by NMFS and industry sources differ depending on the time period used, species considered, and fishing port.

On December 4, 2007, the Senate agreed to S.Res. 376, expressing the sense of the Senate that the Secretary of Commerce should declare a commercial fishery failure for the groundfish fishery for Massachusetts, Maine, New Hampshire, and Rhode Island and immediately propose regulations to implement Section 312(a) of the Magnuson-Stevens Fishery Conservation and Management Act. The Secretary did not change his decision. However, the omnibus spending bill passed on December 17, 2007, included $13.4 million in the NOAA budget for the Massachusetts multispecies fishery. The funding was provided to lessen the economic impacts associated with New England Fishery Management Council's Framework 42 of Amendment 13 to the Multispecies Fishery Management Plan.[11] In August 2008, Massachusetts Governor Patrick announced the disbursement of $11.3 million to Massachusetts fishermen and fishing businesses, $750,000 for crew members, $630,000 for a health insurance program for crew members, and $700,000 to cover administrative fees. Concerns have been raised because fishermen in New Hampshire and Maine who face similar economic hardships are not eligible for Massachusetts funding.

West Coast Salmon Ocean Troll Fishery (Sacramento)

On April 10, 2008, the Pacific Fishery Management Council adopted a complete closure of commercial and sport fisheries off California and most of Oregon in response to the collapse of the Sacramento River fall Chinook salmon run. The minimum conservation goal for Sacramento fall Chinook is 122,000 to 180,000 spawning salmon,[12] while as recently as 2002, 775,000 adults returned to spawn.[13] Even with ocean fishery closures, the 2008 returns of Sacramento fall Chinook were projected to be 54,000 fish. In March 2009, NMFS released a report on the causes of the decline of Sacramento fall Chinook. The report identified unfavorable ocean conditions as the primary factor that led to poor survival of juvenile salmon when they entered the ocean in 2005 and 2006. It also found that the stock was more susceptible to poor ocean conditions because of habitat degradation in the freshwater portion of its range.

On May 1, 2008, in response to requests by the governors of California, Oregon, and Washington, the Secretary of Commerce declared a commercial fishery failure for the West Coast salmon fishery. Congress provided $170 million in disaster funds in the Food, Conservation, and Energy Act of 2008 (P.L. 110-246) for commercial and recreational members of fishing communities who were affected by the fishery failure. In September 2008, $100 million was released to the Pacific States Marine Fishery Commission for distribution to commercial fishermen, processors, charter boat operators, recreational guides, and other businesses dependent on fishing. On April 30, 2009, the Secretary of Commerce notified the governors of California and Oregon that the fishery failure would continue in 2009. Returns of Sacramento fall Chinook salmon remained below levels required for a fishery and the 2009 commercial salmon troll fishery was closed for most of Oregon and all of California. The ocean recreational fisheries were also limited in both states, especially California. The extension of the disaster declaration ensured release of the remaining unspent funds from the original $170 million.

In 2010, revenue from commercial salmon landings in California remained significantly lower than the 2003-2005 average. On September 2, 2010, the Secretary of Commerce continued the fishery failure for California commercial salmon fisheries under Section 3 08(d) of the IFA and 3 12(a) of the MSFCMA.

Gulf of Mexico Fisheries (Hurricanes Gustav and Ike)

On September 17, 2008, Commerce Secretary Gutierrez determined that Hurricanes Gustav and Ike had caused a fishery resource disaster in the Gulf of Mexico. The determination was made pursuant to Section 3 08(d) of the Interjurisdictional Fisheries Act. Commercial harvests in the affected areas of Louisiana and Texas consist mostly of shrimp, finfish, and oysters. In November 2008, NOAA announced that Louisiana had become eligible for up to $40 million and that Texas had become eligible for up to $7 million in fishery disaster aid for restoring damaged oyster reefs, removing storm debris, and rebuilding fishing infrastructure destroyed by the hurricanes.

Chesapeake Bay Blue Crab

On September 23, 2008, Secretary Gutierrez determined that the soft shell blue crab fishery[14] of the Chesapeake Bay had undergone a commercial fishery failure under § 312(a) of the MSFCMA. The blue crab population has declined since the 1990s, with a 41% decline in the value of soft shell blue crab landings in Maryland and Virginia. Although the cause is uncertain, factors contributing to the blue crab population decline are likely to include deteriorating water quality, loss of habitat, and overfishing.[15] Maryland and Virginia adopted new commercial and recreational regulations starting in 2008 to shorten the season in both states, limit the harvest of female crabs in Maryland, and close the winter dredge fishery in Virginia. The fishery failure determination was made in response to requests by the governors of Virginia and Maryland based on the decline of the resource and the importance of this fishery to Chesapeake Bay communities and the regional economy. The federal government allocated $10 million of disaster relief to each state. Funds have been used for assistance to the fishing industry, habitat restoration, and developing permit buyback programs.

New England Red Tide, Puget Sound Sockeye Salmon, and Yukon River Chinook Salmon

On November 14, 2008, the Secretary of Commerce determined a commercial fishery failure had occurred in Maine, New Hampshire, and Massachusetts because a red tide bloom required closure of shellfish fisheries. Blooms of the algae *Alexandrium fundyense*, commonly referred to as red tide, produce a toxin that is ingested and concentrated by shellfish such as clams, mussels, and oysters. When the concentration of the algae is high, shellfish beds must be closed because shellfish become toxic and can cause paralytic shellfish poisoning. During 2008, red tide was widespread in ocean waters off the three states. A total of $5 million was allocated among the three states, with $2 million for Massachusetts and Maine and $1 million for New Hampshire. On December 22, 2010, the Secretary of Commerce determined that red tide caused another fishery failure in the Maine shellfish fishery, but no funding has been appropriated for this case.

On November 14, 2008, the Secretary of Commerce determined a commercial fishery failure had occurred for sockeye salmon fisheries in Puget Sound and the Northern Pacific coast of Washington. Northwest Native American tribes and non-tribal fishermen have been harmed by declines in sockeye salmon runs and harvests. A total of $2 million was provided for tribal and non-tribal fishing communities.

On January 15, 2010, the Secretary of Commerce determined a commercial fishery failure had occurred for the Yukon River king salmon commercial fishery. Harvests were restricted because of the low number of salmon that returned to the Yukon River in 2008 and 2009. In 2008, the commercial fishery harvest was 87% below the recent five-year average, and in 2009 there was no commercial season and limited subsistence fishing. The Supplemental Appropriations Act, 2010 (P.L. 111-212), provided $5 million for fishery failures declared by the Secretary of Commerce in January 2010.[16]

Gulf of Mexico Oil Spill

On May 24, 2010, Secretary of Commerce Locke determined that the ongoing oil spill had caused a fishery failure in the states of Louisiana, Mississippi, and Alabama. On June 2, 2010, the Secretary added Florida to the earlier determination. Both determinations were made under Section 312(a) of the Magnuson-Stevens Fishery Conservation and Management Act (MSFCMA). The Supplemental Appropriations Act of 2010 (P.L. 111-212) included a total of $28 million for fishery disaster assistance and $5 million for economic development assistance. NOAA is expected to allocate $15 million of the fishery disaster assistance for a strategic marketing plan and health and safety assurance program for Gulf coast seafood. The remaining $13 million of fishery disaster assistance is intended to address economic impacts on fishermen and fishery-dependent businesses only if resources provided under other authorities are insufficient. The Supplemental Appropriations Act of 2010 also included $10 million for expanded stock assessments for Gulf fish species and $1 million for a National Academy of Sciences study of the long-term effects of the oil spill. [17]

In response to economic harm caused by the oil spill and to fulfill obligations as a responsible party, BP established a claims process and multiple claims centers.[18] On May 3, 2010, BP began paying emergency compensation to individuals and businesses. BP stated that emergency payments would continue as long as individuals and businesses could show they were unable to earn a living because of injury to natural resources caused by the oil spill. Payments were based on one month of income and would be adjusted with additional documentation. [19] On June 16, 2010, President Obama announced that BP had agreed to set aside $20 billion to pay economic damage claims caused by the oil spill. Some members of the fishing industry criticized BP 's administration of the claims process because they maintained that the claims process was slow, some individual payments were inadequate, and required proof of past earnings might not reflect potential earnings (if the spill had not occurred) for 2010.

On August 23, 2010, the Gulf Coast Claims Facility (GCCF) took over the administration of claims from BP to address the issues with BP 's claims process.[20] Up until the change, BP had paid out $395.6 million, of which approximately $111 million went to the fisheries industry. Although still funded by BP, GCCF was established by the Administration and BP to provide an independent claims process.[21] Some have argued that the process is not independent because the GCCF is directly financed by BP. The GCCF provides information on how to file a claim, how to check the current status of a claim, and other general information concerning the process. The GCCF offered emergency payments equivalent to six months of lost income to claimants until November 23, 2010. As of February 12, 2011, the GCCF had made emergency payments of approximately $751 million to individuals and businesses in the fishing industry.[22] Some claimants have voiced concerns with the transparency of the claims process, the lack of information in GCCF responses to claims, and the adequacy of payments, but the actual proportion of unsatisfied fishing industry claimants is not known at this time.

Recent Administrative Actions by NMFS

On January 15, 2009, NMFS proposed regulations for fishery disaster provisions of § 312(a) and § 315 of the MSFCMA and § 308(b) and § 308(d) of the IFA.[23] The regulations are being developed to clarify these statutes to ensure consistency and to facilitate the processing of requests. NMFS is proposing to establish definitions and characteristics of commercial fishery failures, fishery resource disasters, serious disruptions affecting future production, and harm incurred by fishermen. For example, specific percentage revenue decreases are proposed for commercial fishery failures, and percentage biomass decreases are proposed for serious disruptions affecting future production. The regulations would also establish requirements for initiating NMFS review and the administrative process for processing applications. Final regulations have not been published.

ISSUES FOR CONGRESS

Commercial fisheries are strongly influenced by environmental conditions that may affect industry infrastructure or the abundance and distribution of the fishery resource. These changes often take place suddenly, as in the case of hurricanes, oil spills, and harmful algal blooms within a fishing season with little or no warning. For example, significant portions of federal and state waters in the Gulf of Mexico were closed within the first month of the Deepwater Horizon oil spill. Disaster relief programs may help save businesses that have been harmed by these events and can address severe economic fluctuations by providing assistance until conditions return to "normal." Several concerns have emerged that relate to the nature of commercial fisheries and disaster relief programs, including (1) timing relief to meet crucial needs, (2) relating disaster relief to long-term fisheries management, (3) defining a fishery failure, and (4) determining the beneficiaries of relief.

Timing of Relief

The timeliness of disaster relief is a concern because relief funds are seldom appropriated in anticipation of disasters. First, information regarding the scope of the disaster usually needs to be compiled by the fishing industry, state and local governments, and NMFS. Difficulties in concluding this task can be compounded by the lack of data and readily available economic studies. In cases such as Hurricane Katrina, it was immediately clear that a disaster had occurred, and the Secretary made a determination within two weeks of the landfall of Hurricane Katrina. Although the full dimensions of the disaster and the level and scope of resource needs remained uncertain for months after the disaster, many have asserted that some basic aid should have been provided to members of the fishing industry immediately after the disaster. Similarly, the full extent of damages from the Deepwater Horizon spill is unknown, but closures have already harmed commercial and recreational fishing and related businesses.

For the West Coast troll salmon fishery in 2006, immediate questions revolved around whether a resource disaster would occur. Background information, fishery landings, and economic data were needed for the Secretary to make a determination. Managers and participants were aware of the impending fishery closure before regulations were adopted because the poor condition of the Klamath River Chinook salmon stock was well documented. Even after regulations were adopted, some questioned whether a fishery failure could be declared before the season started and the fishing industry had actually been harmed.

After a fishery failure is declared, funding is dependent on appropriations by Congress. Given the timing of appropriations bills and congressional schedules, it can be difficult to appropriate funding in a timely manner. Hurricane Katrina and Hurricane Rita fishery disaster funding was appropriated in June 2006, more than nine months after the Gulf fishery failure was declared in September 2005. Many in the industry asserted that the greatest need occurred immediately after the hurricanes, when infrastructure, vessels, gear, and markets were lost to fishermen and other industry participants. The West Coast troll salmon fishery was declared a fishery failure in the summer of 2006, but funding was not appropriated until May 2007. Many who work in the Gulf fishing industry are concerned that compensation from BP may not be sufficient or reach them in time to sustain their businesses. Fishery failure determinations have been made for four states in the Gulf of Mexico, and in July 2010, funds for disaster relief to supplement claims or to provide other assistance were approved by Congress.

In the short term, many fishing industry participants believe that the most pressing concern should involve getting relief to those individuals and businesses most directly and immediately affected by the fishery failure. For these needs, some have advocated establishing a disaster fund with funding appropriated in advance that could provide assistance on short notice.

Long-Term Management Approaches

Often direct or indirect assistance to the fishing industry is part of a relief program. Some have criticized federal assistance because it delays the inevitable readjustment that may be needed for fisheries with excess harvesting capacity. Critics argue that climatic and/or environmental conditions are blamed for fish population declines caused by overfishing. Most fish populations vary over time, and frequently it is difficult to determine the relative importance of the factors that cause these variations.

Features of several programs such as buybacks, fisherman training, and cooperative data collection focus on concerns related to the need for readjustments in fishing fleet size. Yet, when relief is provided, even when it includes a buyback program, greater numbers of fishermen and effort usually remain in the fishery than would be sustainable in the long-run. Many fisheries managers agree that relief such as vessel buybacks needs to be more closely integrated with ongoing fisheries management objectives.

Defining Fishery Failures

The general causes of fishery resource disasters that result in commercial fishery failures are defined by the MSFCMA and IFA. However, specific characteristics of a fishery resource disaster such as scale, timing, and extent are not defined. Since there is no set definition of a fishery failure or fishery resource disaster, the Secretary of Commerce has a large degree of discretion when determining whether a fishery failure has occurred.

For example, in 2007 the governors of Maine, Massachusetts, and Rhode Island requested the Secretary of Commerce to declare a fishery failure for the Northeast Multispecies (groundfish) fishery. There appears to be general agreement that Northeast fishermen have faced continuing hardships during the last several years, but disagreement centers on the primary cause of the decline and whether this disruption rises to the level of a commercial fishery failure. These ambiguities appear to be one of the reasons that NMFS solicited the public for information related to fishery resource disasters during 2008 and proposed regulations in January 2009.

Who Benefits?

Who benefits from disaster funding is a reoccurring point of contention. Participants such as fishermen and fish processors may be widely dispersed and difficult to locate and track. Although it is often possible to contact vessel and processing plant owners, industry-related labor such as crew members and fish processing employees may be difficult to track. In some fisheries, crew members are temporary laborers that follow fishing opportunities.[24] Because of the transient nature of employment in the fishing industry and seasonal movement of fishing vessels among regions, labor statistics regarding the employment of fishermen are either difficult to obtain or may not exist. Similar problems may occur in related fishery processing and distribution sectors. Some have voiced the need for better labor statistics that can assist in forecasting and planning for the effects of different fisheries programs, including disaster relief.

Economic effects of fishery disasters on the local community and region are also difficult to quantify. Services directly related to fishing such as boat repairs, dock services, and fishing equipment suppliers, as well as other businesses indirectly related to fishing are likely to be harmed by losses in the fish harvesting and processing sectors. Although general regional impacts can be estimated using economic models, it is often difficult to identify the level of impacts on these businesses because of their dispersed nature and their indirect relationship to fishing. Many have claimed that a broader understanding of these community impacts is needed. Some also argue for more deliberate and long-term data collection and planning to link community concerns with marine fisheries management.

End Notes

1 See the National Oceanic and Atmospheric Administration, Fishery Disaster Assistance website at http://www.nmfs.noaa.gov/mb/financial_services/disaster.htm.

2 NMFS Procedures and Guidance for Disaster Assistance Under Magnuson-Stevens Act Section 312(a) and Interjurisdictional Fisheries Act Sections 3 08(b) and 3 08(d) can be found at http://www.nmfs.noaa.gov/directives/.

3 Eight regional Fishery Management Councils were created by the Fishery Conservation and Management Act, later renamed the Magnuson Fishery Conservation and Management Act and more recently the Magnuson-Stevens Fishery Conservation and Management Act. Council members are appointed by the Secretary of Commerce from lists of candidates knowledgeable of fishery resources, provided by state governors. The councils prepare fishery management plans (FMPs) for those fisheries that occur primarily within the federal waters of the Exclusive Economic Zone (3-200 nautical miles from shore). Links to individual Council websites are available at http://www.nmfs.noaa.gov/.

4 For SBA purposes, disasters may also be declared by the President, state governor, Secretary of Agriculture, or Secretary of Commerce.

5 For information concerning SBA Economic Injury Disaster Loans, see http://www.sba.gov/ services/ disasterassistance/index.html. Also see CRS Report RL33243, *Small Business Administration: A Primer on Programs*, by Oscar R. Gonzales.

6 For information on EDA programs, see http://www.eda.gov/AboutEDA/Programs.xml.

7 National Marine Fisheries Service, Procedures Guidance for Disaster Assistance Under Magnuson-Stevens Act 312(a) and Interjurisdictional Fisheries Act 308(b) and 308(d), National Marine Fisheries Service Instruction [31-108-01], May 8, 2007, p. 10, https://reefshark.nmfs.noaa.gov/f/pds/publicsite/documents/procedures/31-108-01.pdf.

8 See CRS Report 97-441, *Commercial Fishing: Economic Aid and Capacity Reduction*, by Andrew G. Read and Eugene H. Buck.

9 The conservation objectives under the Pacific Fishery Management Council's (PFMC) Pacific Coast Salmon Fishery Management Plan require returns of 33-34% of potential adult natural spawners and no fewer than 35,000 naturally spawning adults to the Klamath River. When the stock is projected to fall below this level, PFMC is required to recommend a

closure of the salmon fisheries within its jurisdiction that harvest Klamath River fall Chinook salmon.

[10] From 2001 to 2005, the dressed weight of Oregon and California troll salmon landings averaged 8.025 million pounds, but in 2006 landings dropped to 1.529 million pounds. For West coast troll salmon fishery statistics, see http://www.pcouncil.org/salmon/salbluebook/salbluebook.html.

[11] The primary purpose of Framework 42 of Amendment 13 to the Multispecies Fishery Management Plan is to establish a biennial adjustment process to review the fishery periodically and recommend changes to management measures necessary to end overfishing and rebuild stocks.

[12] The number of salmon needed to return to the river to sustain this salmon population.

[13] For Pacific salmon fishery management information, see http://www.pcouncil.org/.

[14] Blue crab are harvested at three stages—as hard crab, as peeler crabs (just prior to molting), and as soft shell crabs (immediately after molting).

[15] See http://www.dnr.state.md.us/fisheries/regulations/bluecrabproposedregulations.html

[16] The only disaster declared in January 2010 was for the Yukon River Chinook salmon fishery.

[17] See http://www.commerce.gov/print/news/press-releases/2010/09/17/us-commerce-secretary-gary-locke-announces26-million-gulf-states-fis.

[18] Pursuant to OPA (33 U.S.C. 2713-2715).

[19] U.S. Congress, House Committee on the Judiciary, *Testimony of Darryl White, Vice President, Resources, BP America*, 111th Cong., 2nd sess., May 27, 2010. Hereinafter cited as BP 2010.

[20] See http://www.whitehouse.gov/the-press-office/fact-sheet-claims-and-escrow.

[21] See http://www.gulfcoastclaimsfacility.com/faq#Q1.

[22] See http://www.gulfcoastclaimsfacility.com/reports.

[23] National Oceanic and Atmospheric Administration, "Magnuson-Stevens Act Provisions; Interjurisdictional Fisheries Act; Disaster Assistance Programs; Fisheries Assistance Programs," 74 *Federal Register* 2467-2478, January 15, 2009.

[24] For more information, see CRS Report RS213 12, *How Many Commercial Fishermen?*, by Eugene H. Buck.

In: Impacts of the Gulf Oil Spill on Fishing ... ISBN: 978-1-61470-971-8
Editors: C. Scarnati and E. Popeo

Chapter 5

EFFECTS OF OIL ON MARINE MAMMALS AND SEA TURTLES*

National Oceanic and Atmospheric Administration

Oil is a mixture of chemicals, all of which may have different effects on marine animals and the combination may be even more hazardous. In addition, some of the chemicals and methods used to clean up oil spills may also have effects on marine animals. Toxicity or harmful effects are dependent upon:

- The mixture and types of chemicals that make up the oil or are used to clean up the oil.
- The amount of exposure (dose for internal exposures or time for external exposures).
- The route of exposure (inhaled, ingested, absorbed, or external).
- The biomedical risk factors of the animal (age, sex, reproductive stage, and health status). For turtles this will include differing impacts and vulnerabilities at the different life stages such as eggs, post-hatchlings, juveniles and adults. For cetaceans this will include neonates, calves, juveniles and adults.

Research on dolphins in human care has shown that the animals avoid oil on the surface of the water, however observations of wild dolphins have

---

* This is an edited, reformatted and augmented version of National Oceanic and Atmospheric Administration, U.S. Department of Commerce publication, dated May 12, 2010.

documented the animals swimming in, feeding in and socializing in oiled water during previous oil spills in the Gulf of Mexico.

Several aspects of sea turtles put them at risk, including the lack of avoidance behavior of oiled waters and indiscriminate feeding in convergence zones.

Whales, dolphins, manatees and sea turtles are air breathers and all must come to the surface frequently to take a breath of air. In a large oil spill, these animals may be exposed to volatile chemicals during inhalation.

Depending on the size of a particular spill, marine mammals and sea turtles could be exposed to these chemicals for a fairly long time.

General Types of Effects on Marine Mammals and Sea Turtles

The following are generalities and would need to be specifically applied to the situation or event.

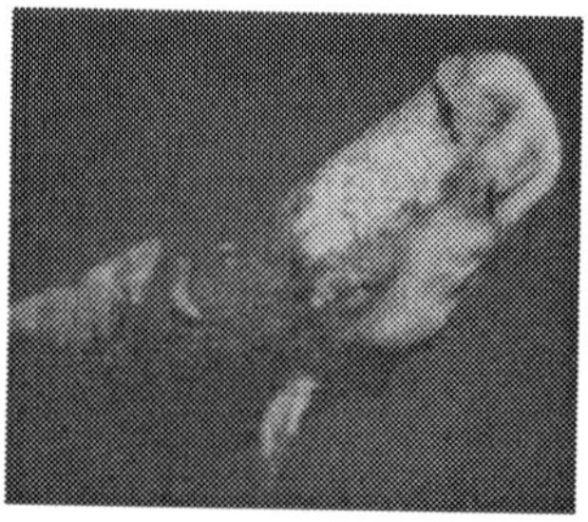

Cetaceans, manatees and sea turtles may be exposed to chemicals in oil (or used to treat oil spills like dispersants) in two ways: internally (eating or swallowing oil, consuming prey containing oil based chemicals, or inhaling of volatile oil related compounds) and externally (swimming in oil or dispersants, or oil or dispersants on skin and body).

Additionally sea turtles may experience oiling impacts on nesting beaches and eggs through chemical exposures resulting in decreased survival to hatching and develop¬mental defects in hatchlings.

EXTERNAL EFFECTS

- Cetaceans and manatees have no fur which can be oiled and do not depend on fur for insulation. Therefore they are not susceptible to the insulation effects (hypothermia) that often puts haired marine mammals (such as fur seals or sea otters) at risk.
- Oil and other chemicals on skin and body may result in skin and eye irritation, burns to mucous membranes of eyes and mouth, and increased susceptibility to infection. For large whales, oil can foul the baleen they use to filter-feed, thereby potentially decreasing their ability to eat.

INTERNAL EFFECTS

- Inhalation of volatile organics from oil or dispersants may result in respiratory irritation, inflammation, emphysema, or pneumonia.
- Ingesting oil or dispersants may result in gastrointestinal inflammation, ulcers, bleeding, diarrhea, and maldigestion.

- Absorption of inhaled and ingested chemicals may damage organs such as the liver or kidney, result in anemia and immune suppression, or lead to reproductive failure or death.

SUMMARY

- Direct contact with of petroleum compounds or dispersants with skin may cause skin irritation, chemical burns, and infections.
- Inhalation of volatile petroleum compounds or dispersants may irritate or injure the respiratory tract, which may lead to inflammation or pneumonia.
- Ingestion of petroleum compounds may cause injury to the gastrointestinal tract, which may affect the animals' ability to absorb or digest foods.
- Absorption of petroleum compounds or dispersants may damage liver, kidney, and brain function as well as causing anemia and immune suppression.
- Long term chronic effects such as decreased survival and lowered reproductive success may occur.
- The Marine Mammal and Sea Turtle Stranding Networks have protocols and procedures in place for responding to live animals that are exposed to oil spills, and animals brought into rehabilitation facilities are provided veterinary care to remove oil and treat any related health effects.

In: Impacts of the Gulf Oil Spill on Fishing … ISBN: 978-1-61470-971-8
Editors: C. Scarnati and E. Popeo

Chapter 6

SEA TURTLE STRANDINGS AND THE DEEPWATER OIL SPILL*

National Oceanic and Atmospheric Administration

Sea turtle stranding responders, working under the guidance of the Wildlife Branch Unit of the Unified Command, which includes several NOAA experts, are responding to dead and live sea turtle strandings and offshore teams are searching for and rescuing oiled sea turtles from the spill area.

All the live oiled turtles are cleaned of oil and are being treated and cared for at one of four primarily de-oiling centers across the northern Gulf of Mexico or at secondary rehabilitation facilities once they are stabilized.

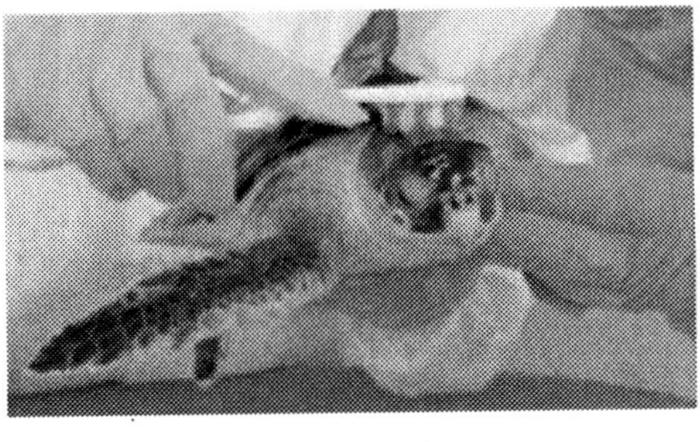

Credit: University of California, Davis

* This is an edited, reformatted and augmented version of National Oceanic and Atmospheric Administration, U.S. Department of Commerce publication, dated July 20, 2010.

NOAA experts are examining turtles that have stranded in the area affected by the oil spill to determine, if possible, whether their deaths can be linked to oil, or another cause. All of the turtles are being sampled externally for oil and tissue samples are being taken during necropsy when the condition of the carcass is sufficiently fresh. The majority of the turtles recovered as strandings have not had external oil.

In contrast, the vast majority of turtles captured offshore during directed surveys are externally oiled. Aerial surveys have documented sea turtles swimming in oiled areas and in unoiled areas. We do believe that this spill will significantly affect sea turtles in the Gulf of Mexico.

General Effects of Oil on Sea Turtles

Sea turtles may be exposed to chemicals in oil or to chemicals in products such as dispersants used in two ways: internally (eating or swallowing oil, consuming prey containing oil based chemicals, or inhaling of volatile oil related compounds) and externally (swimming in oil or dispersants).

Several aspects of sea turtle behavior put them at risk including the importance to turtles of surface convergence areas, typically highly productive areas where ocean currents converge and where oil has been found. These areas provide feeding and sheltering habitat to sea turtles in the Gulf of Mexico, especially the young, early life stage animals.

Sea turtles are air breathers and all must come to the surface frequently to breathe. In an oil spill, they may be exposed to volatile chemicals at the surface during inhalation.

Additionally, sea turtles may experience oiling impacts on nesting beaches when they come ashore to lay their eggs, and their eggs may be exposed during incubation potentially resulting in increased egg mortality and/or possibly developmental defects in hatchlings.

Hatchlings emerging from their nests may encounter oil on the beach and in the water as they begin their lives at sea.

External Effects: Oil and other chemicals on skin and body may result in skin and eye irritation, burns to mucous membranes of eyes and mouth, and increased susceptibility to infection.

Internal Effects: Inhalation of volatile organics from oil or dispersants may result in respiratory irritation, tissue injury, and pneumonia. Ingestion of oil or dispersants may result in gastrointestinal inflammation, ulcers, bleeding, diarrhea, and maldigestion. Absorption of inhaled and ingested chemicals may damage organs such as the liver or kidney, result in anemia and immune suppression, or lead to reproductive failure or death.

REASONS FOR SEA TURTLE STRAND INGS

There are thousands of sea turtle strandings every year along the Gulf of Mexico and U.S. east coast. NOAA is working to understand why sea turtles are stranding in the area of interest. In previous years, this same area has experienced increased strandings during this time of year. The stranding numbers we are seeing currently in Louisiana, Mississippi and Alabama are much higher than normal, however. This may be due in part to increased detection and reporting, but this does not fully account for the increase.

The primary human cause for sea turtle deaths in the Gulf of Mexico is bycatch in fishing gear. Bycatch in shrimp trawls is recognized as a leading source of sea turtle mortality if Turtle Excluder Devices (TEDs) are not properly used or not required.

The other primary types of fishing gear that incidentally catch and can kill sea turtles include longline gear and gillnets. Vessel strikes also cause mortality of sea turtles, especially where turtle abundance and vessel activity are high, such as areas near ports, marinas, and navigation channels.

There are five sea turtle species in the Gulf of Mexico: Kemp's ridley, leatherback, hawksbill, and green sea turtles are listed as endangered under the Endangered Species Act; loggerhead sea turtles are listed as threatened but are currently proposed for endangered listing in the western North Atlantic, which includes the Gulf of Mexico.

REHABILITATING OILED SEA TURTLES

The Wildlife Branch, Marine Mammal and Sea Turtle Unit of the Unified Command has implemented region-wide protocols for caring for turtles in distress and established four rehabilitation/de-oiling centers: one in Louisiana, one in Mississippi, and two in Florida. Additional facilities are on standby if needed.

NOAA is working closely with the U.S. Fish and Wildlife Service, through the Unified Command, on addressing issues of concern on sea turtle nesting beaches in the Gulf of Mexico as well. Wildlife teams on the water have been instructed to bring any turtles in distress back to shore for transport to the rehabilitation facilities.

To get the latest numbers of sea turtle strandings and captures as part of the Wildlife Branch Unit's on water search and rescue operation go to http://response.restoration.noaa.gov/deepwaterhorizon.

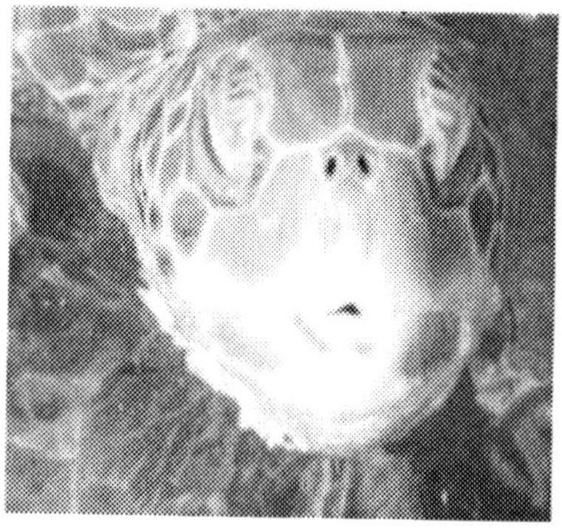

Sea turtles may experience oiling impacts on nesting beaches when they come ashore to lay their eggs. Their eggs may be exposed during incubation potentially resulting in increased egg mortality and/or possibly developmental defects in hatchlings.

In: Impacts of the Gulf Oil Spill on Fishing ... ISBN: 978-1-61470-971-8
Editors: C. Scarnati and E. Popeo

Chapter 7

OIL SPILLS AND CORAL REEFS*

National Oceanic and Atmospheric Administration

Healthy coral reefs are among the most biologically diverse and economically valuable ecosystems on earth, providing valuable and vital ecosystem services.

Coral ecosystems are a source of food for millions; protect coastlines from storms and erosion; provide habitat, spawning and nursery grounds for economically important fish species; provide jobs and income to local economies from fishing, recreation, and tourism; are a source of new medicines, and are hot spots of marine biodiversity. They also are of great cultural importance in many regions around the world, particularly Polynesia.

Based on current estimates, shallow water coral reefs occupy approximately 284,300 square kilometers (110,000 square miles) of the sea floor. If all of the world's shallow water coral reefs were placed side-by-side, they would occupy an area a bit larger than the state of Texas.

The total area of coral reefs represents less than 0.015 percent of the ocean. Yet coral reefs harbor more than one quarter of the ocean's biodiversity. No other ecosystem occupies such a limited area with more life forms.

---

* This is an edited, reformatted and augmented version of National Oceanic and Atmospheric Administration, U.S. Department of Commerce publication, dated May 12, 2010.

Reefs are often compared to rainforests, which are the only other ecosystem that can boast anywhere near the amount of biodiversity found on a reef. Coral reefs are sometimes called rainforests of the seas.

Impacts of oil spills to coral reefs are difficult to predict because each spill presents a unique set of physical, chemical, and biological conditions. How corals are exposed to oil — and the composition of the oil at the time of impact — bears directly on how serious the impact will be.

There are three primary modes of exposure for coral reefs in oil spills:

- Direct oil contact is possible when surface oil is deposited on intertidal corals that live near the surface of the water and become exposed with the tides.
- Rough seas and a light, soluble oil can combine to mix the oil into the water below the surface, where it can impact corals. Corals are exposed to less oil beneath the water surface, but the lighter oil components that mix easily are often the most toxic.
- Subsurface oiling can occur when heavy oils weather, or mix with sediment material. This increases the density of the oil to the point where it may actually sink, potentially smothering corals.

OIL SPILL RESPONSE STRATEGIES FOR CORAL REEFS

Booms are sometimes used to control the movement of oil at the water surface. This should be done carefully in coral reef areas, as boom anchors can physically impact corals, especially when booms are moved around by waves.

Dispersants act like detergents, breaking an oil slick into droplets that mix into the water column, where they dilute and eventually biodegrade. Dispersants work best on light oils, and are less effective on oil that has been extensively weathered or in areas of low water movement. Dispersants offer a trade-off of oil effects in the water versus at the shoreline.

The use of dispersants over shallow submerged reefs is generally not recommended, but the potential impacts to the reef should be weighed against impacts that might occur to birds, mammals, turtles, and sensitive shoreline resources (such as mangroves) where it is extremely difficult to clean the oil.

NOAA has produced two summary documents on corals and oil spills:

A guide for planning and incident response - (http://response.restoration.noaa.gov/book_shelf/70_coral _full_report.pdf)

A synthesis of previous research on oil effects to corals - (http://response.restoration.noaa.gov/ book_shelf/1 _coral_tox.pdf).

In 2005, NOAA conducted an exercise to test emergency response to a simulated oil spill in the Florida Keys (http://sanctuaries.noaa.gov/safeseas/pdfs/safe seas2005.pdf).

EFFECTS OF OIL AND DISPERSANTS ON CORAL REEFS

Laboratory, field studies, and actual oil spill events often appear to show contradictory results for effects of oil and dispersants on coral reefs.

The old notion that coral reefs do not suffer acute toxicity effect from oil floating over them is probably incorrect. Direct contact with spilled oil can lead to coral death, but depends on coral species, growth form, life stage, and type/duration of oil exposure.

Longer exposure to lower levels of oil may kill corals, as well as shorter exposure to higher concentrations. Death may not be immediate, but rather take place long after the exposure has ended.

Instead of acute mortality, it is more likely that oil effects occur in sublethal forms, such as reduced photosynthesis, growth, or reproduction. Early developmental forms (like coral larvae) are particularly sensitive to toxic effects, and oil slicks can significantly reduce larval development and viability.

Coral communities may recover more rapidly from oil exposure alone than from mechanical damage. Recovery of coral reefs after oil exposure, however, may depend partly on the recovery of associated communities (e.g., nursery or foraging habitats, such as mangroves and seagrasses) that may be more seriously affected than the reef itself. Recovery time depends on the type and intensity of the disturbance and can range from several years to decades.

Past Oil Spills Impacts to Coral Reef Ecosystems

One extensively studied spill occurred at Bahia Las Minas, Panama in April, 1986. An estimated 60,000 - 100,000 barrels of medium weight crude oil spilled into the waters of the bay, causing widespread lethal and sub-lethal effects to coral.

In contrast, in the Arabian (Persian) Gulf Spill in January 1 99 1, the largest oil spill in history, an estimated 6.3 million barrels of oil were released. Given the magnitude of this release and the coral reef impacts noted at other tropical spills, there were dire expectations of severe impacts to reefs in Kuwait and Saudi Arabia. However, to date, the extent of coral reef damage directly attributable to the Gulf Spill has been remarkably minor.

What You Can Do

NOAA's comprehensive efforts in response to the Deepwater Horizon event can be found at http://www.incidentnews.gov/incident/8220.

Because oil is a hazardous material, volunteer opportunities are limited for unaffiliated, untrained volunteers. To report oiled shorelines or request volunteer information, please call 281-366-5511 or 866-448-5816.

Learn more about NOAA's response to the BP oil spill at http://response.restoration.noaa.gov/deepwaterhorizon.

To learn more about NOAA, visit http://www.noaa.gov..

INDEX

C

D

E

F

M

N

O

P

R

S

T

U

V

W

Z